Behavioral Ecology of Tropical Birds

Behavioral Ecology of Tropical Birds

Bridget J. M. Stutchbury

and

Eugene S. Morton

ACADEMIC PRESS
A Harcourt Science and Technology Company

San Diego   San Francisco   New York   Boston
London   Sydney   Tokyo

Copyright © 2001 by ACADEMIC PRESS

Academic Press
A Harcourt Science and Technology Company
Harcourt Place, 32 Jamestown Road, London NW1 7BY, UK
http://www.academicpress.com

Academic Press
A Harcourt Science and Technology Company
525 B Street, Suite 1900, San Diego, California 92101-4495, USA
http://www.academicpress.com

ISBN 0-12-675555-8 (hdbk)
0-12-675556-6 (ppbk)

Library of Congress Catalog Number: 00-107415

A catalogue record for this book is available from the British Library

Designed and typeset by Kenneth Burnley
Printed and bound in Great Britain by MPG Books Ltd, Bodmin, Cornwall

01 02 03 04 05 06 MP 9 8 7 6 5 4 3 2 1

Contents

Preface

The idea for this book arose out of necessity. We needed information on mating systems in tropical birds to compare with DNA fingerprinting studies of temperate birds. We were asking a simple question: is extra-pair mating more common in temperate than in tropical passerine birds? We found little information on tropical birds. So we started some empirical field studies in Panama to answer our own question, and found no extra-pair fertilizations in the Dusky Antbird. But we expected to find EPFs in the Clay-colored Robin because we knew from our prior research that they bred synchronously during the dry season, much like temperate zone birds do in the spring. A reviewer of our paper stated that the prediction that Clay-colored Robins should have EPFs qualified us for membership in the Flat Earth Society. Extra-pair mating systems in passerines were (and still are) considered ubiquitous, so it seemed silly to the reviewer that we were making a big issue of predicting that robins would have EPFs. We had come face to face with the Temperate Zone Bias.

Of course this was not the first time. ESM began working in Panama in the 1960s, before behavioral ecology blew on the embers of the dying field of ethology. Early work included latitudinal differences in avian frugivory and fruiting seasons, the influence of nest predation on breeding seasons, and the bioacoustic basis for the evolution of songs in tropical birds. Major differences between temperate and tropical birds were highlighted. We then turned to migratory birds. What a great opportunity they provide to contrast adaptations to differences in latitude within the same individual. But through all these endeavors, it remained our impression that studies of temperate zone birds provided the data to model generalities, and that tropical exceptions were considered oddities. Today, the now vibrant field of behavioral ecology is still much too reliant on these temperate-based models.

There is an intellectual vacuum to fill. We planned this book, not to fill the vacuum, an impossible task, but to stimulate others to work on tropical birds using a new perspective. The new perspective is exciting.

Our premise is not 'why tropical birds are so different' but rather 'why temperate zone birds are so atypical.' Alexander Skutch (1985) used the same logic when he stated that the question should not be 'why do tropical birds lay so few eggs?' but, rather, 'why do temperate zone birds lay so many?' The answer seems more tractable when you ask it in this way.

In the tropics diversity is the name of the game. For example, over 90% of North American passerines have a similar territorial system, they defend breeding territories for only a few months each summer. But, in the tropics, only 13% of passerines defend territories during the breeding season only; instead the predominant territorial system is year-round defense of feeding and nesting territories plus three other systems not represented at all in temperate zone passerines! Our message is clear. In order to discover generalities about avian biology, a diversity of adaptations helps provide the comparative material needed to overcome the thin slice of time represented by the present. And, while understandable, a temperate zone bias is inexcusable, because it is more than a latitudinal bias, it acts as a blinder to the amazing diversity in behavioral adaptations that remain to be explained.

We also have regrets. We apologize for the heavy load we place upon passerine birds in this book. We hope that the ideas are generalizable to other groups. Passerines make good subjects, though, because they are mainly freed from stringent nest site requirements and there are so many species. Our focus on the neotropics is due to our familiarity with the natural history of the birds there.

This familiarity is due largely to the efforts of two mentors, Martin Moynihan and Eugene Eisenmann. Both were instrumental in the development of tropical bird study and in the development of one of the premier tropical research institutions, the Smithsonian Tropical Research Institute (STRI). STRI afforded ESM both predoctoral and postdoctoral opportunities to become familiar with tropical birds and, for both of us, a yearly visit to Panama for research. We thank STRI staff for their help in facilitating our field research, and their excellent library was an invaluable resource for us.

Readers will see, time and again, that we draw conclusions and make generalizations based on evidence from just a few studies and species. For most important questions there are not enough data to perform formal comparative analyses of temperate versus tropical species. Instead we take the few pieces of the puzzle that exist, and our own experience, and try to see the big picture. We cannot wait until dozens of studies have been done on a variety of tropical birds to tackle particular questions. The slow but steady rate at which such studies are being done means that the tropical ecosystem will be largely ruined by the time such comparative studies could be made. But important differences in ecology and behavior do exist, and it is very clear that temperate species are not a good model for understanding the behavioral ecology of tropical birds.

This book is a call to arms. We highlight the missing pieces of the puzzle in the hope that an army of graduate students and researchers will set out to find the answers before it is too late. Our fervent wish is that residents in tropical countries will be stimulated to answer the many questions we raise. Opportunities abound for discovering, describing, and discussing the beautiful ways tropical birds are different from run-of-the-mill temperate zone birds and yet more representative of avian adaptations worldwide.

We thank Isabelle Bisson, Debbie Buehler, Sharon Gill, Gail Fraser, Joan Howlett, Jennifer Nesbitt, Ryan Norris, and Trevor Pitcher for reviewing and commenting on various chapters in this book. The Smithsonian Institution, through its Scholarly Studies Program, and the Natural Sciences and Engineering Research Council of Canada provided essential grant monies to carry out our research and support students. York University provided excellent support for field research by BJMS and her students, and much of this book was written during her sabbatical leave. Stan and Pat Rand provided us with a place to stay and a trusty Cherokee to ride in for several years. We are forever grateful to them. We also thank Douglas and Sarah who were born to the task.

Bridget J. M. Stutchbury
Eugene S. Morton

1 | Why are tropical birds interesting?

1.1 Ecology and breeding seasons

'Tropical birds' brings to mind exotic and showy species like motmots, toucans, parrots, manakins, birds of paradise and groups like the antbirds which are restricted entirely to the neotropics. One pictures a lush humid jungle where these birds thrive year-round, though tropical habitats include savannahs, mangrove and dry forests. The sheer number of different species is overwhelming. It is a simple matter to show a first-time visitor, as we have done in Panama, more than 75 new species of birds in their first morning. A small country like Panama has some 900 species of birds (Ridgely and Gwynne 1989), more than all of North America! Even more impressive because Panama comprises an area (about 75,000 km^2) equivalent to the state of South Carolina. This amazing species diversity is well known to most biologists and naturalists.

Not fully appreciated is the fundamentally different ecology that tropical birds exhibit. Daylength and temperature do not vary seasonally to any great extent. There is no such thing as winter, with short daylight coupled with bitterly cold temperatures. The year is divided into the dry season and wet season not the spring, summer, winter, and fall seasons of the temperate zone. The timing and length of the dry and wet seasons varies with locale and latitude (Figure 1.1). The wet season north of the equator often occurs from May through November, and below the equator at the opposite time of year. Near the equator there are often two dry and wet seasons each year. In some areas there is no distinct dry season (see Forsyth and Miyata 1984, Kricher 1997 for overview). In other areas a prolonged dry season results in a deciduous forest, where some or all trees drop their leaves to reduce water loss. Food is still abundant, however, as nectar and fruit are produced in large supply (Morton 1973, Janzen 1975, Foster 1982, Fleming *et al.* 1987). The year-round availability of food means that most species are residents (non-migratory) and many that eat insects defend territories all year. Intra-tropical and altitudinal migration occur, but are restricted mostly to frugivores (Morton 1977, Levey and Stiles 1992).

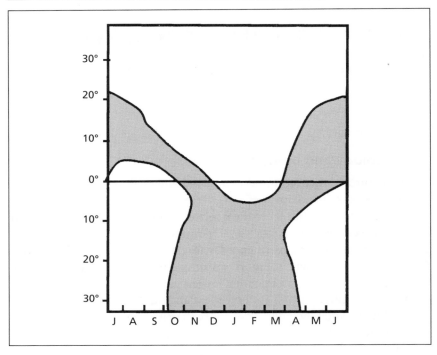

Figure 1.1
Timing of the wet season (shaded) in east Africa at different latitudes and times of year (after Moreau 1950).

Unlike birds of the temperate zone tropical birds breed at all times of the year. Frugivorous birds often breed during the dry season, whereas insectivorous birds breed during the longer wet season (Morton 1971b, Morton 1973). Breeding seasons, typically four to eight months long (Ricklefs 1969b, Kunkel 1974), are timed to coincide with fruit or insect abundance or reduced predation pressure, not climate per se (Chapter 2). This contrasts sharply with the temperate zone where climate is a major constraint that forces birds to breed quickly, within two to three months.

Tropical/temperate zone differences in migratory behavior and breeding season set the stage for major differences in social behavior. This book describes and evaluates the evolutionary consequences of these latitudinal differences as they affect life history traits, mating systems, territoriality and communication. The first simple and broad generalization has already been alluded to: species in temperate regions are under strong selection from abiotic factors (e.g. climate) whereas in tropical regions biotic selection pressures are most important.

Interactions with other species (plant and animal) play a key role in shaping the behavioral adaptations of tropical birds, and are the subject of the final chapter.

1.2 Species diversity

Most biologists identify taxonomic diversity as the greatest difference between temperate zone and tropical birds. Species diversity increases dramatically in the tropics. For instance, there are only 5 genera of tyrant flycatchers in eastern Canada and the U.S., but a remarkable 79 genera in tropical Brazil (Figure 1.2). Similarly for hummingbirds and tanagers (1 versus over 30 genera). Other families like hawks and wrens show a similar but less dramatic pattern (Figure 1.2). Many of our temperate zone birds derive from tropical ancestors. Some groups, such as antbirds, do not occur at all in north temperate areas.

Much tropical research has focused on documenting and understanding patterns of diversity throughout the tropics (Remsen 1984, Terborgh *et al.* 1990, Thiollay 1994), often as part of a strategic biodiversity assessment program. These research efforts have led to the

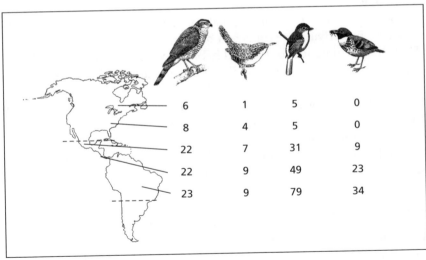

Figure 1.2

Number of genera within each family at different locations in the New World (eastern Canada, southeastern US, Mexico, Panama, and Brazil). Families are Accipitridae (hawks, eagles, kites), Troglodytidae (wrens), Tyrannidae (flycatchers) and Formicariidae (antbirds). The dashed lines indicate the latitudinal boundaries of the tropics (23°N and 23°S). Drawings from Sick (1993), Owings and Morton (1998), Skutch (1997) and Wetmore (1972).

discovery of new species (e.g. Robbins *et al.* 1994, Kennedy *et al.* 1997, Whitney and Alvarez 1998). Part of the mystique of the tropics is the continual discovery of new species, something that rarely (if ever) happens in north temperate regions. Without a doubt, an emphasis on species diversity helps identify areas important for conservation.

We are all too familiar with the shocking facts. Tropical forests are being cut at a rate of some 100,000 ha y^{-1} in the Philippines, 1.5 million ha y^{-1} in Brazil, etc. Worldwide this adds up to 15 million ha y^{-1}. For most people, though, these statistics are too impersonal to really hit home. Anyone who has visited tropical regions can see for themselves, endless miles of landscape of scrubby and often eroded grass where lush tropical forest once stood. Favorite study sites or birding spots that when revisited a few years, or even months or weeks later, are barren of trees. Tropical habitats are being destroyed so quickly that without basic information on which species occur where, and in what numbers, we cannot develop a strategy for saving biodiversity.

But another kind of diversity has been largely neglected in the rush to catalog the occurrence of bird species. Biotic interactions have shaped a behavioral and morphological diversity in tropical birds that is far richer than that found in temperate zone birds. Understanding the behavioral diversity of tropical birds requires that the selection pressures underlying the traits can be inferred from current processes. With the alarming loss of tropical habitats we lose not just the individuals of a given species, but also the ability to study and understand the remarkable adaptations represented through these species. History is being lost. The strong biotic selection pressures mean that disruption of the environment and loss of species can quickly erase the evidence necessary to piece together evolutionary processes in the tropics. Ant-following birds are among the first to disappear from forest fragments, along with members of mixed species flocks (Chapter 7).

1.3   Temperate zone bias in behavioral ecology

It is ironic that tropical birds are viewed as strange, and perhaps even bizarre, when they vastly outnumber temperate zone species (Figure 1.2). About 80% of all passerine species breed in tropical regions. Likewise for other bird groups many of which do not occur outside of the tropics. Temperate zone birds form a minority species group that have converged to adapt to a temperate climate. Because of their worldwide dominance, tropical birds typify the adaptive realm of birds and it is their natural history that should be viewed as the norm for birds.

Shockingly little is known about most tropical birds, even their basic natural history.

Most theory in avian behavioral ecology comes from models and empirical studies of birds in temperate regions. We contend these theories do not apply equally well to tropical birds, because the ecological and social backdrop for tropical birds is fundamentally different. There is a temperate zone bias because the vast majority of biologists are based in temperate regions of North America and Europe, many of whom are ignorant of the unique ecology and behavior of tropical birds. Often these behavioral ecologists and ornithologists do not realize that the conventional wisdom applies only to a select group of birds from temperate regions, birds that do not represent general adaptations of birds.

Several temperate zone species stand out as frequently used models for testing behavioral ecology theory. More behavioral ecology papers have been published on the Red-winged Blackbird, *Agelaius pheoniceus*, (Searcy and Yasukawa 1995, Beletsky 1996) than for all tropical birds combined. One could just as easily substitute the Barn Swallow, *Hirundo rustica* (Møller 1994) or Great Tit, *Parus major*. This is not a criticism of these studies, but a way to put the temperate zone bias in perspective. Yet why shouldn't the Dusky Antbird or some other tropical bird be our model of a typical bird?

Behavioral ecology has not ignored tropical birds or tropical adaptations. The tropics, however, is generally viewed as a place to go to study oddities, or in other words, phenomena that are uncommon in the temperate zone. Many researchers have focused their attention on cooperative breeders (Emlen 1981, Ligon 1981, Rabenold 1990, Komdeur *et al.* 1995, Restrepo and Mondragón 1998) and lekking species (Foster 1981, Trail 1985, McDonald 1989, Westcott 1997), even though these types of social organization do not predominate in the tropics (Kunkel 1974). Other tropical phenomena, like ant-following (Willis 1967, 1972, 1973, Willis and Oniki 1978), mixed-species flocks (Moynihan 1962, Munn and Terborgh 1979, Powell 1985), and duetting behavior (Farabaugh 1982) have similarly been well studied.

Tropical specialties attract attention precisely because they are clearly different from temperate zone systems. The more typical tropical birds are socially monogamous, wherein a male and female defend a territory and raise young together, and defend territories year round (Chapter 5). They may appear the same as their temperate zone counterparts, but recent research has shown that they differ in many ways. The division of labor between members of long-term monogamous associations is nearly equal but has been described for only a few tropical species (e.g. Greenberg and Gradwohl 1983). These typical tropical species have an impressive number of complex and unique adaptations which are only beginning to be described and appreciated. Below we give an overview of several examples where extensive research on temperate zone species has led to broad generalizations about birds, but where tropical birds differ dramatically from these temperate zone systems. This is an overview of more detailed treatment in later chapters.

1.4   Examples of temperate zone bias

Extra-pair mating systems

Over the past decade, genetic testing of parentage (usually via DNA fingerprinting) has spawned many studies revealing that extra-pair fertilizations (EPFs) are common in species previously considered to be monogamous. A female has a single social mate with which she raises young, but a proportion of her young (often over 20%) are derived from copulations with males other than her social mate (Birkhead and Møller 1992). Hence the current distinction between social monogamy (raising young together) versus genetic monogamy (mating exclusively with each other). Although there is much variation in EPF frequency among species (Stutchbury and Morton 1995, Westneat and Sherman 1997), the majority of socially monogamous passerines studied to date have an extra-pair mating system. EPFs drive the evolution of anti-cuckoldry behaviors in males and underlie sex role divergence in reproduction. Extra-pair mating systems are now considered the norm for most birds. Consider the following quote from Birkhead and Møller (1996, p.323) 'Until recently monogamy was also assumed to imply an exclusive mating relationship between two individuals (Wittenberger and Tilson 1980), but recent behavioral and molecular studies (reviewed in Birkhead and Møller 1992) have shattered the illusion of sexual fidelity: in the majority of species extra-pair copulations and fertilizations outside the pair bond occur routinely'.

But are extra-pair mating systems the norm? The majority of parentage studies on socially monogamous birds (over 90%) focus on temperate zone breeders. Male tropical passerines have smaller testes than temperate zone species, often 1/10th the size, which predicts a low level of sperm competition among these males and few EPFs in tropical species (Stutchbury and Morton 1995). The handful of socially monogamous tropical birds fingerprinted have few (< 15%) or no extra-pair young (Chapter 4). Extra-pair matings are likely to be relatively uncommon in socially monogamous tropical birds and, therefore, extra-pair mating systems are not the norm for birds.

This dramatic difference in mating system stems from differences in the length of breeding season which are determined largely by latitude and which cause differences in breeding synchrony between the temperate zone and tropics. Across passerine species, as nesting synchrony increases so does the percentage of broods that contain young derived from EPFs (Chapter 4). The tropics show us that extra-pair mating systems are not typical of passerines, as thought by most biologists. Instead, climatic conditions force temperate zone birds to breed synchronously, and synchronous breeding, in turn, fosters the evolution of extra-pair mating systems (Stutchbury and Morton 1995, Stutchbury 1998a).

Testosterone and territory defense

A high level of testosterone in male birds is thought to be critical for successful territory defense and increasing male attractiveness to females. For example, in socially monogamous species, levels of testosterone are high early in the breeding season during territory establishment and pair formation, but drop when males are feeding young (Wingfield and Moore 1987, Wingfield et al. 1990). Male testis size increases dramatically prior to breeding. In polygynous species, where male parental care is lower and mate attraction persists for most of the season, testosterone levels remain high (Beletsky et al. 1995). When testosterone levels in males are experimentally increased, male parental behavior is suppressed but male attractiveness to females increases (Wingfield 1984, Oring et al. 1989, Ketterson et al. 1992), including success in getting extra-pair fertilizations (Raouf et al. 1997).

Recent studies show that this scenario does not fit tropical birds (Chapter 5). Males of tropical species retain small testes even during the breeding season, and have low levels of circulating testosterone despite having vigorous territory defense and song output. This shows that high testosterone level is not a prerequisite for successful territory defense and mate attraction. Instead, high testosterone in temperate zone birds should be viewed as an adaptation to compete successfully for mates and extra-pair matings during the short temperate breeding season. Thus constrained to nest synchronously, temperate birds evolve specific adaptations to compete within the extra-pair mating systems there, one of which is a crucial reliance on hormonal 'jacking up' based on testosterone.

Territory acquisition

Territory establishment by the male, followed by mate attraction, is the common model of territory defense (Freed 1987). In this kind of territorial system, dominated by long-distance migrants of the temperate zone, males and females have many unoccupied areas for establishing territories when they return in spring. Territory defense against males intruding for extra-pair matings is crucial for males (Stutchbury 1998b). Song, primarily a male trait, coincides with territory establishment, mate attraction and EPF competition. Tropical birds, many of which are year-round residents, face dramatically different opportunities and constraints in acquiring a territory and mate. Year-round territoriality, stable territory boundaries and high adult survivorship results in a low turnover rate on territories (Chapter 5). Extra-pair matings are uncommon, so boundary disputes by males and females are about real estate. Territorial openings occur relatively infrequently so males and females alike have few opportunities to choose mates. Singing occurs at a relatively low rate, often by both sexes, and functions primarily in territory defense rather than mate attraction or EPF competition (Chapter 6).

Although opportunities to switch territories are scarce, individuals are primed to do so. This can be demonstrated by capturing and detaining territory owners for several days to create vacancies experimentally. Our own work on the Dusky Antbird, *Cercomacra tyrannina*, shows that males and females quickly (usually within hours!) abandon mates and territories when given the chance (Morton 1996b, Morton *et al.* 2000). Within minutes of 'losing' their mate, both sexes begin singing a courtship song to attract a new mate. This, despite the fact that Dusky Antbirds often remain with the same mate for five or more years, defend year-round territories, and sing in duets. Similar results have been found for several other tropical passerines where temporary removals have been conducted (Levin 1996a, Gill and Stutchbury, in prep.).

We will expand on the examples above, and others, to convince naturalists and behavioral ecologists that lessons learned from the temperate zone do not necessarily apply in tropical regions. The main theme of this book is to illustrate where, how and why tropical birds are so different from temperate zone birds. The book's purpose is to dispel the temperate zone biologists' ignorance of tropical biology and to stimulate more research on tropical birds. To this end we suggest a theoretical framework based upon latitudinal differences in extra-pair mating and biotic interactions and their influence on life history traits in tropical birds.

2 | Breeding seasons

Interest in what controls the timing of breeding in birds goes back to earlier ornithological studies (Moreau 1937, Nice 1937, Baker 1938), many decades before behavioral ecology was known as a distinct field. In spite of the numerous early studies of breeding seasons, many of which focused on tropical birds (e.g. Skutch 1950, Miller 1962, Ricklefs 1966, Fogden 1972, Sinclair 1978), the question of how selection affects timing of breeding remains a popular and important question in behavioral ecology (Martin 1987, Svensson and Nilsson 1995, Schoech 1996, Svensson 1997). The first, and still most widely accepted, hypothesis is that the availability of food determines when birds breed (Lack 1954, Perrins 1970). The tropics are an especially good place to test this idea because of the great variability among species in diet and their timing of breeding.

2.1 Seasonality in tropical breeding seasons

The tropics are often viewed as rather benign with respect to climate, and since food is generally available year-round, one might expect tropical birds to breed year-round. While it is true that in some species and areas breeding can occur throughout the year (e.g. Miller 1962, Tallman and Tallman 1997), most tropical birds show surprisingly strong seasonality in breeding (Skutch 1950, Sinclair 1978, Boag and Grant 1984). There are distinct and more or less predictable times of the year when a species does not breed. Two general patterns distinguish tropical breeding seasons from temperate zone ones: (1) greater length of breeding season in the tropics, and (2) greater variability among species (and individuals) in when breeding occurs.

Tropical species are characterized by breeding seasons that are two to three times as long as those typical for the temperate zone, and this applies to a variety of bird groups worldwide (Baker 1938, Skutch 1950, Ricklefs 1966, Wyndham 1986). For example, averaging breeding season length for a large number of species (mostly passerines), Ricklefs (1966) found breeding season length to range from 3.1

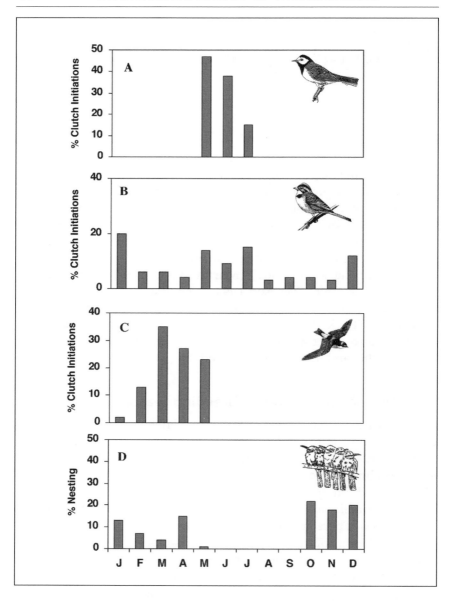

Figure 2.1

Breeding season length for A) Typical temperate zone passerine, the Hooded Warbler, *Wilsonia citrina* (Evans Ogden and Stutchbury 1996) B) Rufous-collared Sparrow, *Zonotrichia capensis* (Miller 1962), C) Mangrove Swallow, *Tachycineta albilinea* (Moore et al. 1999) and D) White-fronted Bee-eater, *Merops bullokoides* (Wrege and Emlen 1991). Drawings from Owings and Morton (1998), Wetmore (1984), and Krebs and Davies (1991).

to 4.2 months in the temperate zone, and 6.6 to 9.8 months in tropical regions. In a broad review, Baker (1938) found a similar latitudinal pattern in breeding season for avian groups such as seabirds, herons, ducks and raptors.

North temperate species of landbirds generally breed at the same time, May–July (Figure 2.1). Short breeding seasons in temperate zone species clearly result from climatic constraints; there is only a short window of opportunity where temperatures and food supply allow successful breeding. In temperate regions, there is little variation among species and individuals in when breeding occurs. Most passerines breed in the spring and early summer, and species differ by only a matter of weeks in when breeding is initiated (Lack 1950). Within species, most individuals lay their first clutch within a few weeks of each other. Detailed studies on Blue Tits, *Parus caeruleus*, and Great Tits, *Parus major*, examine the adaptive significance of differences of only several weeks in clutch initiations (Perrins 1991, Nager and van Noordwijk 1995, Ramsay and Houston 1997).

Their tropical congeners have much longer breeding seasons, which vary in time of year from species to species (Figure 2.1). Breeding seasons of different species and individuals are often separated by months rather than weeks. All kinds of patterns can be found in the tropics. Some species breed primarily during the dry season months and others during the wet season. As the timing and length of the dry season changes with latitude, so do the breeding seasons (Snow 1976a). Seasonality is often more pronounced at high altitudes where breeding seasons are shorter (Skutch 1950). In some areas where there are two wet seasons, species show two peaks of breeding activity during the year (Miller 1962, Wilkinson 1983). The breeding seasons of individuals within a species may vary greatly (e.g. Robinson *et al.* 2000) raising the question of why some individuals begin breeding months before others.

Extreme differences in breeding season can also occur over very short distances (Wrege and Emlen 1991). In montane areas of western Cameroon, lowland populations breed 5–6 months later than conspecifics in higher altitude populations only tens of kilometers away (Tye 1991). Clay-colored Robin, *Turdus grayi*, populations in Panama separated by only 30 km breed several months apart (Morton 1973; Figure 2.4). The question, then, is: can food availability predict the timing of breeding in tropical birds?

2.2   Food availability and timing of breeding

The food availability hypothesis suggests, in its simplest form, that birds should breed when food is abundant for raising young (Lack 1954). This should be especially important for birds with altricial young that require extensive provisioning by parents. Perrins (1970) argued that food for producing eggs may constrain timing of breeding because females that lay too early can pay a high price if food abundance is low. In the temperate zone food abundance changes drastically in spring over a few weeks, so there is likely a tradeoff between producing eggs at the best time versus hatching young at the best time.

The food availability hypothesis has been generally supported in the temperate zone. In particular, detailed studies on European tits have shown that egg-laying is timed so that the nestling period coincides with a 2–3 week period of caterpillar abundance in spring. Individuals that lay early or late (naturally, or due to experimental manipulations) suffer increased nestling mortality (Perrins 1991, van Noordwijk *et al.* 1995). Blue Tits supplemented with food advance egg-laying significantly, but by less than a week (Nilsson and Svensson 1993). Mainland and Mediterranean Blue Tits at a similar latitude differ by three weeks in the onset of laying in response to consistent differences in the availability of caterpillars (Zandt *et al.* 1990). This adaptive behavior has a genetic basis and results mainly from differences in responsiveness to similar daylengths (Lambrechts *et al.* 1996). These kinds of details, or anything close to it, are not available for any tropical species.

A simple prediction of the food availability hypothesis is that diet should explain the breeding seasons of different tropical species. Frugivorous species should breed when fruit is abundant, nectarivorous species when flowers are abundant, and insectivorous species when arthropods are most abundant. Examples where breeding seasons generally correlate with rainfall or food abundance are numerous. In Central America, Skutch (1950) reported that nectarivorous birds (hummingbirds) breed during the dry season (January–March), which is when flowers are most abundant. In contrast, birds that eat grass seeds (e.g. *Sporophila*) breed only later in the year (June–August), well after the wet season begins and enough time has passed for the grass to form new leafy shoots, flower and begin to set seed. In a tropical African savannah, the peak in nesting activity of insectivorous birds occurs from December through June, which is when rainfall and insect abundance is high (Sinclair 1978). Little or no reproduction occurs during the dry season months (July–October). These sorts of broad comparisons, which lump many species together into ecological groups, provide only a weak correlation between breeding seasons and food supply at the individual level. Formal comparative studies, which take phylogeny into account, have yet to be done.

A more convincing test would be to show that peaks in breeding activity within a given species coincide with peaks in food abundance. Relatively few studies have actually measured food abundance and correlated this with clutch initiations, and these have provided mixed results. Opportunistic breeding in Darwin's Finches, *Geospiza*, beginning soon after major rainfall, has been interpreted as support for the food availability hypothesis (Boag and Grant 1984). But this does not explain why some individuals begin breeding before the rainfall, and long after food abundance has declined (Boag and Grant 1984). In a hummingbird, the Long-tailed Hermit, *Phaethornis superciliosis*, there is a bimodal pattern of nesting activity which corresponds roughly to the two annual peaks in flower availability (Figure 2.2; Stiles 1980). Nectarivorous Hawaiian honeycreepers, the Apapane, *Himatione sanguinea*, and Iiwi, *Vestaria coccinea*, also have peak breeding activity when flower abundance peaks (Ralph and Fancy 1994). In the White-crowned Pigeon, *Columbia leucocephala*, nesting was closely related to the abundance of one particular fruiting plant (Bancroft *et al.* 2000).

But there are also many examples where food peaks do not coincide closely with breeding. In two species of manakins (*Pipra mentalis* and *Manacus candei*), both of which are almost entirely frugivorous, breeding peaked when fruit abundance was lowest (Figure 2.2; Levey 1988). In several insectivorous species, breeding peaks before arthropod abundance peaks (Figure 2.2; Gradwohl and Greenberg 1982, Young 1994, Komdeur 1996). Similarly, peaks in breeding often precede peaks in rainfall (and presumably food abundance) by several months in other insectivorous birds (Wunderle 1982, Wilkinson 1983, Woodall 1994). It is paradoxical that the peak period for breeding occurs several months before food availability peaks, and such findings certainly seem inconsistent with the food availability hypothesis.

A resolution to this apparent contradiction may come, in part, with a more precise definition of how food availability affects individual fitness. Frugivorous and nectarivorous species likely depend on high-protein insect food for feeding their young (Morton 1973), so this may explain why in some species, like the manakins, breeding occurs when insects, rather than fruit, are at the peak abundance (Levey 1988). But why should insectivorous species breed when arthropod abundance is low? In tropical species where clutch size is low (Chapter 3) and

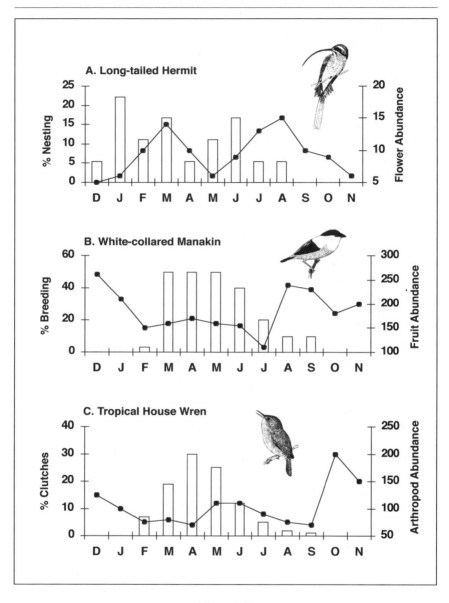

Figure 2.2
Timing of breeding and food abundance for A) Long-tailed Hermit, *Phaethornis superciliosus* (Stiles 1980; % nests started and number of foodplants in full bloom), B) White-collared Manakin, *Manacus candei* (Levey 1988; % individuals captured in breeding condition and total number plants with ripe fruit) and C) Tropical House Wren, *Troglodytes aedon* (Young 1994; % clutch initiations and arthropod biomass). Drawings from Blake (1953).

breeding vacancies scarce (see Chapter 5), reproductive success may be limited by the survival and successful dispersal of fledglings, rather than by the ability of parents to feed a large brood of nestlings in the nest. Fledglings often remain on their parent's territory for many months after they leave the nest, and this period may be crucial for their survival. In Tropical House Wrens, *Troglodytes aedon*, arthropod abundance was highest at the time of juvenile dispersal and molt (Young 1994) suggesting that juvenile survival may be dependent on abundant food.

To understand the selective forces at work, one needs to study the consequences of early versus late breeding on the reproductive success of individuals within a population. In other words, to assess the fitness consequences of different reproductive tactics. Although this approach is common in temperate zone studies (Perrins 1991, van Noordwijk *et al.* 1995), few studies on tropical species have attempted to measure the costs and benefits of breeding at different times. Do individuals that nest very early, or very late, suffer in terms of offspring production? Do the fledgling house wrens produced by early nesting pairs in February (Figure 2.2) have a lower survival or dispersal success because insect abundance is still low when they fledge? Reproductive success in Seychelles Warblers, *Acrocephalus sechellensis*, is highest for nests begun two months prior to the peak in food abundance (Figure 2.3; Komdeur 1996). Two months is the period required for nest construction, laying

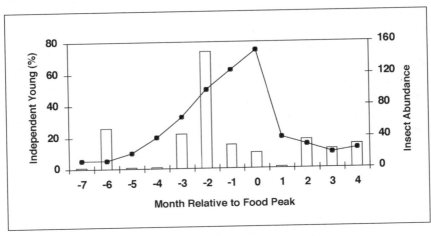

Figure 2.3

Figure 2.3. Percentage of Seychelles Warbler, *Acrocephalus sechellensis*, clutches producing independent young in nests begun at different times of the year (open bars) relative to the peak insect abundance (solid line) which usually occurs some time from July–September. Data from Komdeur (1996).

and incubation, so most pairs have nestlings at the time of peak food abundance.

Experimental manipulations of food supply are one way to determine how selection is operating on individuals. This approach has been used extensively for temperate zone species, where dozens of studies have shown that food supplementation affects timing of breeding by advancing egg-laying dates (reviewed in Martin 1987, Svensson and Nilsson 1995, Schoech 1996). Such experiments have rarely been conducted on tropical species; in fact, we know of only one example. In an African eagle, *Aquila wahlbergi*, food supplementation did not induce earlier laying (Simmons 1993). Komdeur (1996) did manipulate food supply in the Seychelles Warbler, not through food supplementation but through translocation of breeding pairs to islands with differing food supply. Pairs transferred to islands with higher food abundance had prolonged breeding seasons and higher annual reproductive success, compared with their own breeding histories prior to the transfer. Given the very broad variation in timing of breeding among individuals in some populations, food supplementation experiments have the potential for dramatic effects.

2.3   Nest predation and molt

While the food availability hypothesis is strongly supported with detailed experimental studies in the temperate zone, there is much conflicting evidence from studies on tropical birds. There are many examples where tropical birds do not breed at a time when food for producing eggs or feeding young is most abundant. These temperate zone hypotheses (Lack 1954, Perrins 1970) explain breeding seasons that are constrained by the temperate zone climate. Other factors must be considered to explain tropical breeding seasons such as nest predation, molt, and sexual selection.

Nest predation on tropical birds can be very high (80–90% of nests, Chapter 3), and predation risk often varies seasonally. Morton (1971b) suggested that Clay-colored Robins breed at a time when food is scarce in order to avoid seasonal peaks in nest predation. Avoidance of nest predation may also explain why Bananaquits, *Coereba flaveola*, begin breeding long before the wet season begins (Wunderle 1982). Food is likely scarce during the dry season as clutch sizes and nestling weights in large broods are lower for nests begun before the wet season. However, nest predation on Bananaquits increases from 30% in the dry season to 72% in the wet season.

Nest predation does not explain all examples of early breeding by insectivorous species. In House Wrens nest predation risk does not vary seasonally (Young 1994). In Dot-winged Antwrens, *Microrhopias quixensis*, predation risk was much higher on nests begun early in the wet season (Gradwohl and Greenberg 1982). Since the only successful nests were those begun late in the wet season, early breeding cannot be explained as an escape from nest predation.

The timing of breeding may be determined by selection on the timing of molt. Molt is energetically expensive, and there is much evidence that tropical species avoid molt at times of year when food is scarce (Fogden 1972, Poulin *et al.* 1992). Breeding and molt generally overlap little or not at all for individuals within a breeding population (Levey and Stiles 1994, Ralph and Fancy 1994, Tallman and Tallman 1997) or within the avian community as a whole (reviewed in Foster 1975). Most species have a distinct molt season that follows breeding, and the molt is more regular in its timing than the breeding season (e.g. Snow 1962, 1974, Fogden 1972, Levey and Stiles 1994). It has been argued that this regularity in the timing of molt means that molt is 'fixed' in its timing which in turn constrains the timing of breeding (Snow 1962, 1974).

Species where individuals extensively overlap breeding and molt generally have a very protracted molt (6–9 months), which likely reduces the costs of overlapping two energetically expensive activities (Stiles and Wolf 1974, Wilkinson 1983, Levey and Stiles 1994). For instance, in the Long-tailed Hermit individuals differ by as much as 6 months in the initiation of molt (Stiles and Wolf 1974). Some males display on leks while in full molt, while others have no overlap between displaying and molting. Remarkably, individuals molt at the same time (± 2 weeks) each year.

However, there is no direct evidence indicating that molt determines breeding seasons. While molt could force birds to end breeding while food is still abundant, it does not explain why breeding begins when it does. As Levey and Stiles (1994) note, the effect of molt on breeding seasons remains poorly understood and studies following marked individuals over several seasons are needed to determine the consequences of different molt/breeding strategies. No studies have examined whether individuals who overlap breeding and molt are at any selective disadvantage.

2.4   Sexual selection and the breeding season of the Clay-colored Robin

We mentioned earlier that the food availability hypothesis may or may not operate in the tropics. Here we take a deeper look at an example where food does not control the breeding season. Instead, the breeding season may be molded by sexual selection to a greater extent than by natural selection. In Panama, the breeding season of a tropical songbird, the Clay-colored Robin, has characteristics that are difficult to explain by natural selection alone. The first is that robins begin to breed at different times in places separated by very short distances (Figure 2.4). The breeding season may begin a month earlier on the

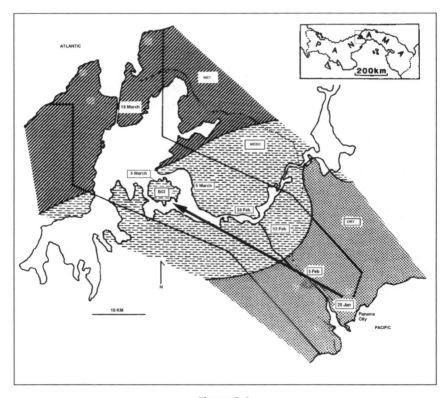

Figure 2.4
Timing of the start of the breeding season of Clay-colored Robins, *Turdus grayi*, in the canal area of Panama (Morton, unpubl. data). Differently shaded areas indicate dry, mesic or wet regions. Nestlings born in Panama City, but translocated to BCI (arrow) and hand-raised there bred at the same time as the nearest local birds on the mainland.

Pacific coast, near Panama City, than 20 km inland at Summit Gardens. Overall, breeding begins in January on the Pacific coast and spreads slowly to the Atlantic, where, while only 80 km away, breeding might not begin until mid-April. Such differences in the timing of breeding over short horizontal distances are unheard of in the temperate zone.

What factors might underlie the timing of breeding in robins? One idea we examined is that breeding seasons are controlled genetically and each population's breeding differs owing to underlying genetic differences. Another idea focuses on food availability, which varies with the beginning of the dry season. The dry season begins earlier and lasts longest on the Pacific than on the Atlantic coast, which accords with the robin's timing of breeding. We did an experimental translocation to test whether genetics or local environment determined timing of breeding. We handraised baby robins (n = 25) taken from the Pacific coast and placed them in large flight cages on Barro Colorado Island (BCI) in Gatun Lake, about halfway between the two coasts (Figure 2.4). BCI is entirely forested and was without wild robins. The nearest population of robins was found in the town of Frijoles, about 3.2 km away over water and well out of earshot of our caged birds. The captives were fed food ad libitum, so food availability was not a factor. For several years (1970–1974) these birds initiated breeding at the same time (first week of March) as the wild birds at Frijoles and about four weeks later than the Pacific coast population from which they came.

The captive robin data rule out either food availability or genetic differences to explain their breeding time. Using natural selection, we suggested that, instead, predation might affect the breeding season (Morton 1971b). Predation risk jumps from 58% of nests destroyed during the dry season to 85% during the wet season, so predation replaces nestling food abundance as the primary determinant of the breeding season. Food availability for nestlings is lowest during the robins' breeding season and most nests suffer severe brood reduction through starvation of nestlings (Chapter 3).

Clay-colored Robins often fledge at very low body mass, basically as runts (see also Chapter 3). Unlike many passerines, they continue to grow in size long after they are independent of their parents. One consequence of this adaptation is that the flight feathers they have as juveniles end up being too small for their growing body. All captive-raised birds molted their wing and tail feathers, in addition to the body feathers, during the postjuvenal molt. Most tropical birds do not molt flight feathers in the postjuvenal molt. The replacement of tail and flight feathers by young robins must be an adaptation to their starvation dry season diet. The nestling robin is caught in the 'altricial strategy'. It cannot eat fruit because fruit will not enable it to grow at the maximum rate needed to escape nest predation during the time its body heat is provided by the mother. It must have high protein food for that. Nestling robins will not eat fruit until they begin to regulate their own body temperature (Morton 1973). As a consequence, brood reduction through starvation is common, even though nestlings are fed fruit much more so than temperate zone *Turdus* (Dyrcz 1983).

Another peculiar feature of robin breeding seasons is that the dawn chorus of males and nesting activity begins abruptly within a population (Stutchbury *et al.* 1998). Even within a population, there are areas where individuals are highly synchronized with each other. In our study site at Gamboa, Panama, for 3 years in a row 5 adjacent males along one street began singing before any others, and the last group of males began singing about 3 weeks later. Small cadres of males, perhaps they could be termed leks (Wagner 1993), begin to sing nearly simultaneously, producing even higher synchrony than found in the town as a whole.

So, even if we are satisfied that predator pressure pushes robins into breeding when food for nestlings is lowest, we still cannot explain another characteristic, their high breeding synchrony within any one location (Stutchbury *et al.* 1998). None of the potential causes from natural selection can explain both breeding at the worst time for raising young and the high localized synchrony in breeding in this tropical thrush.

Sexual selection provides an explanation for both. Females control who mates with them. Females are larger than males in this robin (Morton 1983). In robins, it is likely that male song output in the dawn chorus, after males have been fasting during the night, might be the currency females use in choosing males (Stutchbury *et al.* 1998). By breeding synchronously, females choose males after assessing them under the same ecological conditions (see Chapter 4). In other words, breeding synchrony produces an even playing field upon which males compete. Female choice, we suggest, produced the synchronous onset of singing and nesting. One advantage to breeding in the dry season is that males can recuperate from their singing marathons because food for them is in greatest abundance. This food is fruit, especially *Miconia argentea*, *Xylopia* sp., *Bersera simaruba*, and *Panax morototoni* in our study area, all of which are abundant only in the dry season.

Breeding when ecological conditions favor male song output may evolve if males that sing more mate more. In omnivores like robins, breeding seasons may not be the best time for feeding nestlings but the best time for quickly eating fruit and then returning to singing. In this way, sexual selection can influence the timing of breeding seasons. Other species with strong sexual selection, particularly those with classical leks (manakins, cotingas) may also have breeding seasons that cannot be explained entirely by natural selection. Once again, tropical birds show that the temperate zone data stating that birds breed when it is best for raising young, is not necessarily the case.

Although food availability for making eggs and feeding nestlings is paramount for temperate species, this does not appear to be generally true for tropical birds in terms of their breeding seasons. Food availability fine-tunes breeding seasons in some species, but in others predation or sexual selection is more important.

2.5 Proximate cues

Despite the long history of studies on tropical breeding seasons, little is known about the proximate cues that stimulate individuals to become physiologically prepared to breed. The proximate mechanisms that have evolved can give us insight into the ultimate factors that favor breeding at a particular time. The great variability in breeding seasons among species and individuals in the same locale, and from year to year, suggests that short-term cues (rainfall, food availability, etc.) must trigger gonadal growth.

While photoperiod clearly is the main cue used by temperate zone birds, this cue has long been assumed as unimportant for tropical birds because daylength varies so little near the equator. Ironically, this is an instance where tropical birds are not so different. A recent study on a neotropical forest passerine, the Spotted Antbird, *Hylophylax naevioides*, showed that individuals can perceive the small one hour differences in daylength that occur over the year in its natural habitat (Hau *et al.* 1998). Individuals even responded physiologically to a photoperiod increase of only 17 minutes. In the wild, gonadal growth began 1–2 months prior to the wet season, presumably in response to photoperiod, but short-term cues (rainfall, food) are responsible for the fine-tuning of the start of breeding (Wikelski *et al.* 2000). If rainfall is a cue, then we learn only that birds respond to indicators of the wet and dry seasons and this does not address, at the ultimate level, whether food availability, predation or other factors are important.

Despite these recent discoveries in the physiological and ecological proximate factors that control timing of breeding, we still know nothing about the fitness consequences of different individual strategies. There is great variation within a population in breeding strategies, even though one can presume that the proximate cues experienced (daylength, rainfall) are virtually identical (Figures 2.2, 2.3). In Song Wrens, *Cyphorinus phaeocephalus*, many pairs begin breeding in May, at the start of the wet season, but other pairs did not lay their first eggs until September or October (Robinson *et al.* 2000). The unpredictability of tropical breeding seasons at the individual level is illustrated nicely by White-fronted Bee-eaters, *Merops bullockoides*, in Kenya (Wrege and Emlen 1991). Breeding colonies separated by only a few kilometers breed 6 months apart, during either the long rainy season (March–May) or the short rainy season (October–December). Within a colony, neither insect abundance nor rainfall consistently correspond with time of breeding. Even more puzzling, nests initiated during either the short rainy season fledge three times as many young as long rainy season nests. There is no obvious adaptive explanation for why adjacent colonies breed at different times of the year, but there must be one. Hypotheses developed to explain temperate zone breeding systems are inadequate for explaining this kind of tropical phenomenon.

3 | Life history traits

It has long been recognized that tropical birds differ fundamentally from temperate zone birds in their life history traits. Tropical birds have high nest predation, high adult survival and small clutch sizes (Lack 1947, 1948, 1968, Ricklefs 1969b, Fogden 1972, Skutch 1949). These characteristics in turn have a big impact on the evolution of other behaviors such as mate choice and territory acquisition. More recent studies, however, have questioned the validity of these differences in tropical birds (Karr *et al.* 1990, Martin 1996, Geffen and Yom-Tov 2000), causing some confusion and doubt as to whether tropical birds differ importantly in life history traits. The purpose of this chapter is to review these debates and determine what life history traits characterize tropical birds. Others have carefully reviewed the evolutionary hypotheses to explain why tropical birds are different (Klomp 1970, Murray 1985, Skutch 1985), so this chapter will summarize what is known rather than attempt a comprehensive review.

3.1 High nest predation

Early studies of tropical birds typically reported a high percentage of nests lost to predators, in the order of 80% or more (Snow 1962, Willis 1967, 1972, Fogden 1972, Snow and Snow 1973). In contrast, a predation frequency of 40–60% is typical of many temperate zone songbirds (Martin 1993). Some have argued that high nest predation rates in tropical birds are an artifact of habitat, because a number of the key studies were done in human-disturbed habitats or islands where predation rates may be elevated (Oniki 1979, Martin 1996). But recent studies in large mainland tracts have also found low nesting success (Robinson *et al.* 2000). Other studies that question the high nest predation rate in the tropics have used artificial nests, and found nest losses in the order of 10–50% (Loiselle and Hopps 1983, Gibbs 1991, Sieving 1992). Artificial nests often do not reflect true predation frequency (e.g. Wilson and Brittingham 1998), so these alone cannot be used as evidence for low nest predation in the tropics.

Relatively few studies provide detailed data for nest predation frequency for a large sample of nests of a particular species. A wide diversity of passerines often lose at least 70% of nests (Figure 3.1). This also applies to many non-passerines, like the Rufous-breasted Hermit, *Glaucis hirsuta* (Snow and Snow 1973) and Plain Ground Dove, *Columbina passerina* (Oniki 1979). Robinson *et al.* (2000) found that only 29% of open-cup nesting forest birds in Panama fledged young. Predation is the primary cause of nest failure (Ricklefs 1969b). The percentage of nests lost underestimates nest predation, because this does not take into account when the nest was first found (many early nests could have been depredated and therefore never found). Several studies also used Mayfield's method which estimates daily mortality rate (Young 1994, Roper and Goldstein 1997, Woodworth 1997, Robinson *et al.* 2000). In Dusky Antbirds, *Cercomacra tyrannina*, only 8% of pairs (15/197) raised young to independence over an eight-year study, indicating that nesting success must be very low (Morton and Stutchbury 2000).

Martin (1996) notes some exceptions, tropical species with high nesting success, but these studies were based on relatively small sample sizes and are not comparable (Snow and Snow 1963, Skutch 1981).

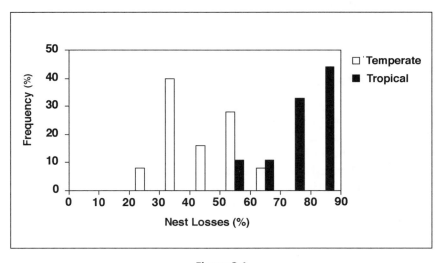

Figure 3.1
Frequency distribution of predation frequency on nests for studies on north temperate passerines (n = 25, Martin 1993) and Neotropical passerines (n = 9; Snow 1962, Morton 1971b and unpubl. data, Willis 1974, Oniki 1979, Wunderle 1982, Skutch 1985, Young 1994, Roper and Goldstein 1997, Woodworth 1997). Only studies with at least 100 nests monitored were included.

Nest predation in the tropics likely varies with habitat (Marchant 1960), time of year (Morton 1971b) and possibly altitude (Skutch 1985). Although very high nest predation (> 80% nests) has been reported for several temperate zone studies (Snow and Snow 1963, Martin 1993), this is certainly not the norm except in highly disturbed habitats (e.g. Robinson *et al.* 1995). For temperate zone passerines the frequency of nest predation averaged 43.7% (Martin 1993), much lower than for tropical passerines (Figure 3.1). Robinson *et al.* (2000), using a more detailed data set, found that open cup nesting temperate zone birds averaged 47% nest loss compared with 71% for tropical birds.

While this sort of comparison is very convincing, many of the tropical species are members of groups that do not have temperate zone counterparts (antbirds, manakins). Many features of behavior and life history could influence predation frequency, so a search for real latitudinal differences owing to habitat should take phylogeny into account. This is not yet possible because so few tropical species have been studied in sufficient detail to estimate predation frequency.

A formal comparative analysis not withstanding, we can conclude that nest predation is higher for most tropical birds. Why is nest predation so high? It is generally assumed that there is a higher number and diversity of nest predators in the tropics. Skutch (1949, 1985) suggested that snakes are the primary nest predators, but other studies do not support this (Roper and Goldstein 1997). Instead, a high abundance and diversity of small mammals, such as mouse opossums, *Marmosa* sp., in the neotropics, may be implicated as the main predator species (Roper and Goldstein 1997).

3.2 High adult survival

Snow (1962) was one of the first to show high annual survival (70%) of adults in a tropical bird, the White-bearded Manakin, *Manacus manacus*. Fogden (1972) did not study any one species intensively, but reported that 200 of 286 (86%) banded adults of a variety of species were alive one year later. Willis (1974) reported survival rates of 69–81% for three species of antbirds in Panama, despite two of the species declining significantly over the study. Most long-term intensive studies of populations report high adult survival based on resightings and recaptures of breeders (Table 3.1, reviewed in Sandercock *et al.* 2000). Such high annual survival rates result in lifespans greater than 10 years being common for these small birds (Snow and Lill 1974,

Table 3.1

Examples of annual survival of territorial adults from population studies of tropical birds. Superscript '*' indicates survival estimates from recapture data of known age individuals.

Species	Years	N	Survival (%)
White-bearded Manakin[a]	9	182	82%
Checker-throated Antwren[b]	14	40*	75%
Slaty Antshrike[c]	7	50	54%
Dusky Antbird[d]	8	25*	82%
Spotted Antbird[e]	10	>100	81%
Long-tailed Hermit[f]	4	105	52%
Medium Ground Finch[g]	16	284	78%
Cactus Ground Finch[h]	16	210	81%
Hawaii Akepa[i]	5	82	79%
Long-tailed Manakin[j]	8	46	78%
Green-rumped Parrotlet[k]	10	>500	68%

a: *Manacus manacus* (Snow 1962, Snow and Lill 1974); b: *Myrmotherula fulviventris* (Greenberg and Gradwohl 1997); c: *Thamnophilus atrinucha* (Greenberg and Gradwohl 1986); d: *Cercomacra tyrannina* (Morton and Stutchbury 2000); e: *Hylophylax naevioides* (Willis 1974); f: *Phaethornis superciliosis* (Stiles 1992); g: *Geospiza fortis* (Grant and Grant 1992); h: *Geospiza scandens* (Grant and Grant 1992); i: *Loxops coccineus* (Lepson and Freed 1995); j: *Chiroxiphia linearis* (McDonald 1993); k: *Forpus passerinus* (Sandercock et al. 2000).

Grant and Grant 1992). Willis (1983) recorded three male Spotted Antbirds, *Hylophylax naevioides*, over 13 years old. Male Long-tailed Manakins, *Chiroxiphia linearis*, do not even begin copulating with females on the lek until they are nine years old (McDonald 1993).

In surprising contrast, Karr *et al.* (1990) used capture-recapture data and Jolly-Seber models to estimate the annual survival of tropical species to be only 56% (n = 25 species), and concluded they did not live longer than temperate zone passerines. As noted by Karr *et al.* (1990) and others (Martin 1996, Greenberg and Gradwohl 1997, Johnston *et al.* 1997, Ricklefs 1997, Sandercock *et al.* 2000), Karr's estimates are not comparable to those obtained from breeding birds in long-term studies. Rather than monitoring known populations, Karr's data for tropical birds come from routine mist netting of a wide variety of species in a given locale. Survival estimates are for all banded individuals regardless of age or territorial status. High dispersal by juveniles, or the presence of floaters, would result in an underestimate of true survival (Lepson and Freed 1995). So would territory switching by adults, common in some tropical birds with year-long territories (Morton *et al.* 2001). This is nicely illustrated in the Green-rumped

Parrotlet, *Forpus passerinus* (Sandercock *et al.* 2000), where survival estimates for non-territorial floaters are much lower than for territorial breeders (Figure 3.2). For many questions in behavioral ecology, we are interested in how long a breeder is likely to survive. This affects decisions about reproductive effort, mate choice and switching territories. Juvenile survival is important too, but for a different set of evolutionary questions such as how to acquire territories, and when and if to disperse.

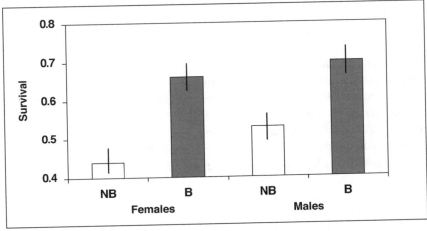

Figure 3.2

Estimates of survival rate (± 1 SE) of adult female (n = 485) and male (n = 849) Green-rumped Parrotlets (*Forpus passerinus*). Nonbreeders (NB) were individuals that did not have a nest cavity but were present on the study site, and breeders (B) were individuals that initiated a clutch. Data from Sandercock *et al.* (2000).

Johnston *et al.* (1997) estimated the annual survival of tropical forest passerines in a long-term mist-netting study in Trinidad to be 65% (n = 17 species). They used analytical models to remove any bias caused by young and transient birds in their sample. Even then, they suggest they have likely underestimated survival. Faaborg and Arendt (1995) found a survival rate of 68% using capture-recapture data for 9 Puerto Rican passerines. Johnston *et al.* (1997) used a linear contrast comparative method to control for phylogenetic effects, and found that tropical species have a significantly higher annual survival rate than comparable temperate zone birds.

No matter which method of survival estimation is used (long-term monitoring of individuals versus capture-recapture), tropical birds average higher survival than comparable temperate zone birds.

A powerful test is to compare survival in a genus that occurs in both temperate and tropical regions (unlike manakins, antbirds and Hawaiian honeycreepers). Ricklefs (1997) used museum collections to estimate survival of New World *Turdus* thrushes to be higher in tropical (0.76–0.85) than north temperate (0.56) species.

3.3   Small clutch size

Tropical birds do have smaller clutches than temperate zone birds, the data are unequivocal (Figure 3.3; Cody 1966, Skutch 1985, Kulesza 1990). The prevailing clutch size is 2 eggs for tropical passerines of the humid neotropics, larger for hole-nesters (Skutch 1985). Within taxonomic groups, clutch size increases two to three fold from the tropics to high northern latitudes where clutches of 4–6 eggs are common (Cody 1966, Klomp 1970). In the southern hemisphere, the relationship is weaker or does not exist (Rowley and Russell 1991, Yom-Tov *et al.* 1994). There are exceptions of course. Some taxonomic groups (Procellariiforms, hummingbirds, pigeons) do not have latitudinal variation in clutch size, and some (gannets, crossbills, ravens) show the reverse trend (reviewed in Klomp 1970). Some tropical birds have an unusually large clutch, like the Yellow-throated Euphonia, *Euphonia hirundinacea*, which lays 5 eggs (Sargent 1993) and the Green-rumped Parrotlet which lays an average of 7 eggs (Beissinger and Waltman 1991).

While the pattern of small clutches in the tropics is real and uncontested, the evolutionary explanation for this temperate–tropical difference has been the subject of long debate (Lack 1947, 1948, Skutch 1949). It is convenient to group these hypotheses into explanations based on immediate costs that limit clutch size versus future costs. Immediate costs result from tradeoffs between clutch size and survival of those offspring, and include reduced food delivery to large broods (Lack 1947, 1948, 1968), increased risk of predation on large broods (Skutch 1949) and a lower likelihood of juvenile recruitment with large broods (Young 1996). Future costs are tradeoffs between current reproductive output and future fecundity and survival, and will be dealt with in the next section. Both kinds of costs can act to limit clutch size.

The food limitation hypothesis suggests that latitudinal differences in daylength allow temperate zone birds to gather more food per day, thus allowing parents more energy to produce eggs and feed altricial young (reviewed in Lack 1947, Klomp 1970, Murray 1985). But it is more than just latitude that affects feeding rates. Growth rates of

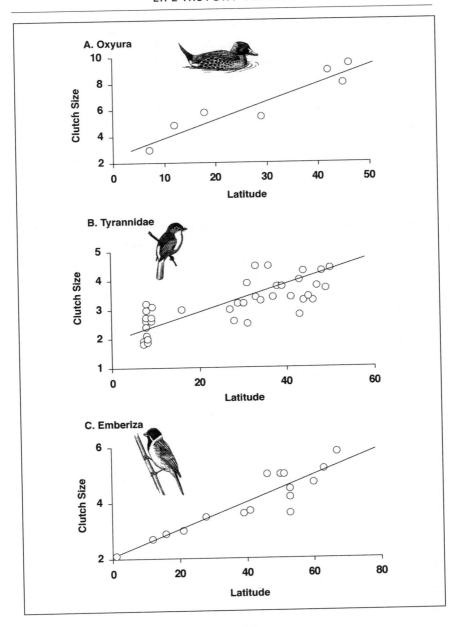

Figure 3.3
Clutch size versus latitude for A) the genus *Oxyura* (stiff-tailed ducks), worldwide
B) the family Tyrannidae (flycatchers), Central and North America C) the genus
Emberiza (sparrows, finches), Africa, Europe and Asia. Data from Cody (1966).
Drawings from Sick (1993), Skutch (1997), and Etchecopar and Hue (1967).

nestlings in the tropics tend to be slower than for comparable temperate zone species (Ricklefs 1969b, 1976) and overall food delivery rates are low. Few studies have been done with reasonable sample sizes, but feeding rates of 1.3–2.2 trips per hour per nestling, which is only about 5 trips h^{-1} to the nest by both parents, are typical in tropical passerines (Greenberg and Gradwohl 1983, Willis 1967, Skutch 1996). This is much less than that of typical temperate zone passerines where the parents may make 20–30 trips h^{-1} to the nest for older nestlings (e.g. Evans Ogden and Stutchbury 1997). Feeding rates in the temperate zone are very high, and they have more hours of daylight in which to keep this up.

Shorter daylength is not the only possible mechanism by which tropical birds are food limited. Although a lush tropical forest brings to mind superabundant food, this is not the case. Although insects and fruit are present year round, insect availability does not reach the amazing peaks seen in the temperate zone summer (see Figure 7.1). Tropical insectivorous birds likely have lower prey availability, regardless of daylength (Janzen 1973, Hails 1982). For instance, insectivorous tropical passerines in French Guiana averaged an attack rate of < 1 per minute, compared to passerines in temperate forests of France who attacked prey at a rate of 2–4 per minute (Thiollay 1988). This is due in large part to an absolute lower prey abundance in French Guiana, particularly for caterpillar density (see also Figure 7.1).

Brood manipulation studies are the standard tool for testing whether food availability limits clutch size. The key predictions are that feeding trips per young, and offspring survival, decrease in experimentally enlarged broods. Temperate zone studies have generally shown that parents can raise experimentally enlarged broods (reviewed in Vanderwerf 1992), suggesting that food limitation does not fully explain clutch size evolution, though it may set an upper limit to clutch size.

In contrast, our experiments with the Clay-colored Robin, *Turdus grayi*, highlight the high frequency of starvation in nests of some tropical birds. In 1970–71, one or two young were added or taken from nests (n = 16) early in the nestling stage during the dry season (Feb–Apr) in Panama. Only one pair could raise a brood larger than the normal 1–3 young (Figure 3.4), and this was the only nest where brood enlargement was done at the beginning of the wet season when invertebrate food for young is more abundant. Brood reduction was due to starvation. The third and smallest nestling either died in the nest (n = 6) or fledged at very low mass (36.4 g ± 9.2 (SD), n = 6) compared with the other nestlings in the brood (largest: 52.1 ± 4.0 g, middle: 45.4 ±

5.9 g, n = 12). Even in unmanipulated nests pairs had difficulty raising three young, averaging only 2.1 fledglings per nest (Figure 3.4).

Brood manipulations in 5 of 7 tropical species found that parents cannot easily raise larger broods (Table 3.2), supporting the food availability hypothesis. These studies found that parents rarely could fledge the enlarged brood, and young in enlarged broods often grew slower or

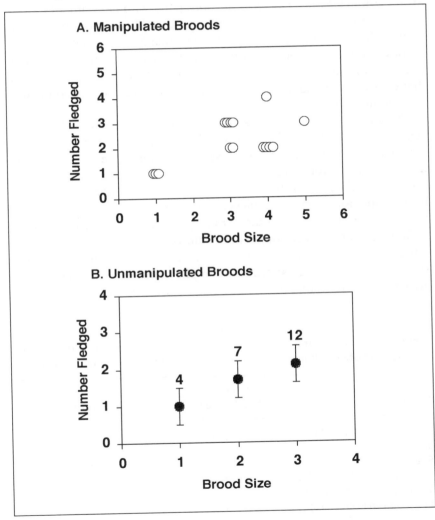

Figure 3.4
Results of brood manipulations in the Clay-colored Robin. Number fledged versus brood size for A) brood manipulations (n = 16 nests) and B) unmanipulated nests (mean ± SE, sample size above).

fledged at lower weights than in broods of normal size. Slow attrition in nests suggests the main cause of nest losses was starvation, rather than predation. There were exceptions. The Tropical House Wren, *Troglodytes aedon*, showed low fledging success of enlarged broods in only one of three years, and number of young fledged averaged higher for enlarged broods than broods of typical size (Young 1996). In Swallow-tailed Gulls, *Creagrus furcatus*, pairs with experimental nests of twins raised twice as many young as normal broods of one (Harris 1970). Stoleson and Beissinger (1997) manipulated hatching synchrony of Green-rumped Parrotlets, which have unusually large clutch size, and found that pairs could successfully feed broods of 8 young that were the same age, suggesting that parents were not constrained in their ability to feed young.

The inability of some tropical birds to raise larger than normal broods can be interpreted several ways. First, there may be a real food limitation that selects for small clutch sizes. Second, feeding rates and nest sizes may be adapted for small clutches to the point that unnaturally large broods are doomed. Lill (1974) added a third nestling to manakin broods, and noted that since the nest was so small the 'extra' nestling had literally to sit on top of the others. Finally, parents may be quite capable of finding food for extra young, but be unwilling to increase their reproductive effort owing to future costs to their own productivity and/or survival (Nur 1990). Despite these complications, brood manipulations are a first step in testing the food availability hypothesis.

Table 3.2

Summary of experimental enlargement of brood size in tropical birds, showing number of broods manipulated (n) and effect on reproductive success.

Species	Enlarged broods fledged more young?	Raised unnaturally large broods?
Swallow-tailed Gull[a] (30)	yes	yes
Savannah Hawk[b] (10)	–	rarely
Snail Kite[c] (23)	no	rarely
White-rumped Swiftlet[d] (27)	no	rarely
White-bearded Manakin[e] (5)	–	no
Clay-colored Robin[f] (16)	no	rarely
House Wren[g] (107)	yes	yes

a: *Creagrus furcatus* (Harris 1970); b: *Buteogallus meridionalis* (Mader 1982): c: *Rostrhamus sociabilis* (Beissinger 1990); d: *Aerodramus spodiopygius* (Tarburton 1987); e: *Manacus manacus* (Lill 1974); f: *Turdus grayi* (Morton unpubl.); g: *Troglodytes aedon* (Young 1996).

Food availability is a function not only of the amount of prey in the environment, but also the degree of competition for those prey. A relatively low degree of seasonality in the tropics may result in populations staying near carrying capacity, and increased competition for food resources means fewer resources per capita (Ashmole 1963, Ricklefs 1980). In the temperate zone, where most species experience high annual mortality, lower densities at the beginning of the breeding season result in a high availability of food per pair. Ashmole's hypothesis predicts similar clutch size among species using different food resources, so long as the degree of seasonality is similar. Lack and Moreau (1965) tested this hypothesis by comparing equatorial birds in different habitats. Savannah species experiencing strong seasonality had higher clutch sizes than forest birds. Marchant (1960) found similar results in Ecuador.

While brood manipulations appear to support the notion that food availability limits clutch size in the tropics, several authors (reviewed in Klomp 1970, Murray 1985, Skutch 1985) have pointed out inconsistent patterns. First, clutch sizes are the same for closely related species with very divergent habitats and feeding ecology. As Skutch (1985) notes, neotropical flycatchers are an ecologically diverse group yet most species have an average clutch size of two eggs. Second, species with biparental care should be able to raise more young than those with uniparental care, but clutch sizes are usually the same regardless of the parental care system (Skutch 1985). Third, nocturnal species also show an increase in clutch size with latitude even though night length (and foraging time) decrease with latitude. Any general hypothesis for clutch size in the tropics must account for these patterns.

Skutch (1949) suggested that nest predation, not food limitation, was the main cost of having large broods. He observed that nest predation is remarkably high in the tropics, and suggested that large broods would suffer high predation owing to the increased parental activity at the nest. This hypothesis assumes that predators find nests primarily via parental activity. There are few direct tests of this idea. Young (1996) found that the feeding rate to House Wren nests increased with enlarged broods but the risk of nest predation did not. Interestingly, this species has a comparably high feeding rate (4–6 trips/h/nestling), perhaps because it is a cavity nester and can afford to do so. Roper and Goldstein (1997) found that for Slaty Antshrike, *Thamnophilus atrinucha*, predation risk was not higher during the nestling stage than incubation stage, despite a much higher visitation rate to the nest. The observation that predation on eggs is as frequent as predation on nestlings (Ricklefs 1969b) is not a valid reason to reject Skutch's hypothesis. Predators, especially nocturnal ones, clearly can find nests without using parental activity as a cue. But, this does not mean that increased parental activity at the nestling stage would not attract nest predators and increase predation risk. There are other mechanisms by which nest predation could favor small clutch size, for instance if this allows birds more energy for renesting after frequent nest failure (Foster 1974).

Less widely recognized is the pattern for many tropical passerines to lay eggs every other day, rather than daily. Most antbirds, flycatchers and manakins in the neotropics skip one day between laying eggs while most oscines lay daily (data from species accounts in Skutch 1960, 1981). While this might suggest the trait is phylogenetically conserved, flycatchers in temperate regions lay one egg each day as do most passerines. The longer period taken to lay eggs would presumably expose the nest to a higher risk of predation. Low food or nutrient availability might explain the longer period for egg-laying, but does not explain why skipping a day during laying is common in the non-oscines but rare in the oscines.

Neither food availability or nest predation can fully explain clutch size patterns in the tropics. A third possible cost of raising large clutches may be the reduced success of juvenile establishment on breeding territories (Young 1996). This is a quantity versus quality argument. Many tropical species are resident and territorial, and the high adult survivorship should result in relatively few breeding vacancies for juveniles. Juveniles from smaller broods may fledge at higher weights, or receive better post-fledging parental care and, therefore, be more likely to obtain to breeding territories (Young 1996). There is no evidence that directly supports this idea, other than studies on cooperatively breeding species which have good information on territory establishment by young birds (see Chapter 5).

3.4   Life history tradeoffs

Large clutch size may be costly not just in the immediate sense, but also in terms of future costs to survival and fecundity. Experiments with temperate zone birds have found that clutch size is often smaller than that which would raise the most young (Vanderwerf 1992). Life history theory offers an explanation: low reproductive effort increases longevity. Reproductive effort is the energy required to produce eggs, defend nests, feed young, etc. Reproductive effort is evolutionarily linked to adult survival through the cost of reproduction. Cost of reproduction is the reduced future survival or reproductive output which results from current reproductive effort. High reproductive effort causes a high cost of reproduction through the physiological toll of breeding, and can also include increased risk of adult predation while breeding (Lima 1987). The key evidence supporting a cost of reproduction comes from brood manipulation experiments with temperate zone birds (Nur 1988, 1990). Individuals with experimentally increased reproductive effort often (but not always) suffer reduced future survival and/or reproductive output compared with individuals raising fewer young.

Few studies have attempted to measure the cost of reproducing in tropical birds. Tropical House Wrens whose brood was increased by two young laid fewer eggs later that year and the next breeding season, but did not suffer decreased survival (Young 1996). Green-rumped Parrotlet females could raise large broods that were manipulated to hatch synchronously without any decrease in adult survival or future reproductive success (Stoleson and Beissinger 1997). Both these species have relatively large clutch sizes, so perhaps the costs of reproduction would be higher for species with the more typical clutch of two or three eggs.

Small clutch size and long lifespan occur in tropical birds, but why? Small clutch size could evolve owing to food shortage and/or high predation, and this low reproductive effort in turn could result in long lifespan. Alternatively, relatively stable tropical environments could result in high adult survival which in turn favors a small clutch size because populations are near their carrying capacity and there is intense competition among young for breeding opportunities (Cody 1966). There is no agreement in the literature (Ricklefs 1977, 1997, Martin 1995).

Do tropical birds have a long lifespan because they lay a small clutch? While this is intuitively appealing, lower clutch size by itself does not necessarily result in lower reproductive effort or a lower cost of reproduction. While it may be reasonable to assume that larger clutch size (and broods) increases the cost of reproduction *within* a species, this is not true for comparisons between species (Lessells 1991). Does a Hooded Warbler, *Wilsonia citrina*, raising four eggs in northeastern United States have a higher cost of reproduction than a Dusky Antbird raising only two eggs in Panama? The future costs to the parent of raising young will depend on a constellation of traits, including food availability for producing eggs and feeding young, clutch size, degree of male parental care (incubation, feeding young), degree of nest defense, duration of incubation, nestling and fledgling stages, effects of breeding on costs of molt and migration, and number of nesting attempts per season. Many, if not all, of these traits differ between temperate and tropical birds. A Dusky Antbird may indeed have to work harder to raise only two eggs (and pay a higher price for it) because food is more limiting in the tropics, renesting is more frequent (owing to high predation), incubation and nestling periods are longer, and the fledged young remain dependent on their parents for many months. However, Dusky Antbirds are resident and do not face intense male–male competition for extra-pair matings (Fleischer *et al.* 1997) which is a large component of reproductive effort in many temperate zone species like Hooded Warblers (Stutchbury *et al.* 1997). Comparing the costs of reproduction between Hooded Warblers and Dusky Antbirds is like comparing apples and bananas. Tropical birds do not have low reproductive effort just because clutch size is smaller.

There is good evidence for immediate (rather than future) tradeoffs: low food availability for young, high nest predation, and limited opportunities for young to obtain good territories all favor small clutch size. Unlike temperate zone birds, tropical birds appear to lay as many eggs as they can successfully raise. The tropical environment that results in high adult survival also favors small clutch size. A cost of reproduction is not necessary, or sufficient, for explaining small clutch sizes in long-lived tropical birds.

3.5   Similarities between tropical and south temperate life history traits

Tropical birds are not the only ones with a long lifespan, high predation rate and small clutch size. Birds in south temperate regions also have this suite of life history traits (Robinson 1990, Rowley and Russell 1991, Yom-Tov *et al.* 1994, Martin 1996, Martin *et al.* 2000) as a result of selection from similar ecological and social pressures. Climate in south temperate regions, like southern South America and Australia, is not nearly as seasonal as it is in north temperate regions because land masses are concentrated at latitudes closer to the equator. Consequently, breeding seasons are long and most passerines are year-round residents (Rowley and Russell 1991). These similarities with the tropics mean that hypotheses for the evolution of life history traits in the tropics can also be tested in south temperate regions, if these hypotheses truly offer general explanations.

Martin *et al.* (2000) tested Skutch's hypothesis that small clutch size evolves in response to high predation, because predators find nests by observing parental visits. They compared nesting success and parental behavior of south temperate passerines with those from North America. For several warbler species in Arizona, greater parental activity during incubation was positively related with actual predation costs. Similarly, within each region, species with high feeding rates to young had higher daily nest predation rates. Thus it seems that Skutch was right about the selective mechanism, in that high predation can constrain parental activity at the nest and result in lower overall food delivery rates. This in turn could set a limit to clutch size. But there is a catch. Although clutch size was much lower in Argentina than Arizona, parents in Argentina delivered food to nestlings at a *higher* rate and did not experience higher nest predation compared with their north temperate counterparts. So while Skutch's hypothesis is supported within each region, it does not explain why the south temperate species have such small clutch sizes. But Skutch's hypothesis may still apply to comparisons between north temperate regions and tropical ones. This is because nest predation generally is much higher in the tropics than the north temperate zone (Section 3.1) and, visitation rates, though poorly studied, seem to be much lower in the tropics (1–2 trips/h/nestling; Greenberg and Gradwohl 1983, Willis 1967, Skutch 1996) than Martin *et al.* (2000) found for their south temperate birds (2–11 trips/h/nestling).

It has become popular to question the validity of tropical–temperate comparisons of life history traits (e.g. Karr *et al.* 1990, Martin 1996, Geffen and Yom-Tov 2000, Robinson *et al.* 2000). This has led to some confusion in the literature. For instance, some authors continue to suggest that adult survivorship is not higher in the tropics (Geffen and Yom-Tov 2000) based on Karr *et al.* (1990). Numerous studies following the Karr *et al.* (1990) paper have shown that it should be interpreted with caution at best (Greenberg and Gradwohl 1997, Johnston *et al.* 1997, Ricklefs 1997, Sandercock *et al.* 2000). Difference in tropical/temperate rates of nest predation has also been questioned (Martin 1996). But again, recent empirical studies (Robinson *et al.* 2000) have supported the early population studies that reported high nest losses in the tropics. The 'dogma' will survive because these patterns have strong empirical support, but it is true that the many exceptions, like those noted by Martin *et al.* (2000), provide the variation needed to fully test hypotheses. At present, experiments are too few to understand why these patterns and variations exist. For instance, in both the tropics and south temperate regions we know very little about food limitation and its effects on parental behavior and clutch size. Brood manipulations, though rarely done, have yielded mixed results; food is clearly not limiting for some species (Young 1996, Stoleson and Beissinger 1997) but others cannot raise additional young (Table 3.2). Martin *et al.* (2000) favor a 'cost of reproduction' explanation for small clutch sizes in south temperate regions, where smaller clutches result in longer lifespans. Although the existence of a cost of reproduction is almost taken for granted, the few studies that have looked for such costs in tropical birds have not found any (Young 1996, Stoleson and Beissinger 1997). The outcome of such experiments in the temperate zone is almost a foregone conclusion, but this is not true for the tropics where we cannot easily anticipate what these experiments will reveal.

4 | Mating systems

4.1 Monogamy and extra-pair mating

The vast majority of birds are socially monogamous, where one male and female form a pair bond and raise young together (Lack 1968). One of the most exciting recent discoveries has been the realization that many of these species are not actually genetically monogamous. DNA fingerprinting has shown that many young, often 20–50%, in monogamous species, particularly passerines, are the result of extra-pair fertilizations, which we will call EPFs for short (see reviews in Birkhead and Møller 1992, Westneat and Sherman 1997, Birkhead 1998). The occurrence of EPFs means, of course, that males often raise young to which they are not genetically related. Thus the social mating system of monogamous pair bonds and shared parental care of young is very different from the genetic mating system, which is promiscuous. Males in monogamous species can potentially fertilize many additional females without forming pair bonds with them. Females are active participants in the pursuit of EPFs. The benefits to females are still being assessed.

The extra-pair mating system of socially monogamous birds has many similarities to leks (Wagner 1993, 1998) because sexual selection is strong and females obtain only sperm through copulations with extra-pair partners. The successful attainment of EPFs is not random or merely opportunistic. To obtain extra-pair copulations (EPCs), males and females have evolved specialized extra-pair mating strategies, like covert visits to neighboring territories and female advertisement of fertility, to compete for extra-pair mates (e.g. Hoi 1997, Stutchbury 1998b, Neudorf et al. 1997). The pursuit of EPCs can be costly for fighting between resident and intruder is common. While extra-pair mating systems are the norm for temperate zone passerines, this is not true for tropical birds.

The first evidence to suggest that tropical birds are genetically as well as socially monogamous came from a comparison of testes size (Stutchbury and Morton 1995). Testes size can be used as an indirect measure of the intensity of sperm competition, and correlates well with EPF frequency (Møller and Briskie 1995). Temperate zone passerines with abundant EPCs and intense sperm competition have much larger testes than tropical passerines (Figure 4.1). For instance, temperate zone Tree Swallows, *Tachycineta bicolor*, have abundant EPFs and testes ten times larger than their close relative in the tropics, the Mangrove Swallow, *T. albilinea*, in which EPFs are much less common (Moore *et al.* 1999). Similarly, even during the peak of the breeding season, tropical Spotted Antbird, *Hylophylax naevioides*, males have testes that are one tenth the size of most temperate zone passerines (Wikelski *et al.* 2000).

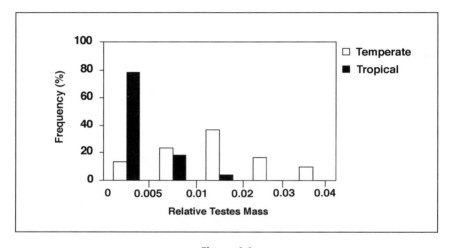

Figure 4.1
Testes mass (relative to body mass) of Neotropical passerines and North American temperate migrants (data from Stutchbury and Morton 1995).

Relatively few DNA fingerprinting studies have been done on monogamous tropical birds. But, of the studies thus far, tropical species have been found to have few EPFs (Table 4.1). The exception thus far is the Clay-colored Robin, *Turdus grayi*, but this was expected because the breeding ecology of this species is similar to temperate zone species in important ways (see below).

Why are extra-pair matings absent or infrequent in tropical passerines but abundant in the temperate zone? The two most common explanations given for genetic monogamy in general are effective paternity guards and mate retaliation. Frequent copulation and/or close mate guarding by males (close following of fertile females) could override female extra-pair mating tactics and result in no EPFs (Fleischer *et al.* 1994, Mauck *et al.* 1995, Negro *et al.* 1996, Petren *et al.* 1999). Alternatively, if mates of promiscuous females retaliate by

Table 4.1
Frequency of extra-pair fertilizations and degree of breeding synchrony in socially monogamous tropical birds. Values give % of extra-pair young and broods that contained at least one extra-pair young (sample size in parentheses), and the breeding synchrony index (Kempenaers 1993).

Species	EPF frequency		Breeding synchrony (%)
	Young	Broods	
Dusky Antbird[a]	0% (15)	0% (12)	8
Mangrove Swallow[b]	15% (98)	26% (30)	8
Clay-colored Robin[c]	38% (37)	53% (19)	25
Palila[d]	0% (20)	0% (12)	low
Cactus Finch[e]	8% (159)	15% (66)	low
White-eye[f]	0% (122)	0%	12
Green-rumped parrotlet[g]	8% (827)	14% (160)	low

a: *Cercomacra tyrannina* (Fleischer *et al.* 1997); b: *Tachycineta albilinea* (Moore *et al.* 1999); c: *Turdus grayi* (Stutchbury *et al.* 1998); d: *Loxiodes bailleui* (Fleischer *et al.* 1994); e: *Geospiza scandens* (Petren *et al.* 1999); f: *Zosterops lateralis* (Robertson 1996); g: *Forpus passerinus* (Melland 2000).

withholding important parental care, it may not pay for females to seek EPFs in the first place (Birkhead and Møller 1996). But recent experimental studies have shown that paternity guards are not effective because females seek EPCs, and males that mate guard are making the best of a bad situation (Kempenaers *et al.* 1995, Wagner *et al.* 1996). Similarly, cuckolded males generally do not withhold parental care (Whittingham *et al.* 1992a, MacDougall-Shackleton and Robertson 1998) so females can obtain EPFs without paying any price in terms of reduced male care. Besides, neither of the general explanations for genetic monogamy would explain the latitudinal difference in extra-pair behavior. Furthermore, the earlier views stressed that asynchronous nesting would favor extra-pair behavior because males are freed from mate guarding when their social mates have completed egg-laying (Birkhead and Biggins 1987, Westneat *et al.* 1990). This predicts that EPFs should be much more prevalent in the tropics than in temperate latitudes, the opposite of what is known.

We need to look for underlying ecological factors that differ between these regions and that affect the evolution of extra-pair mating systems. The answer to our question 'why so few EPFs in tropical birds?' lies in the dramatic differences in breeding seasons (Chapter 2). Breeding synchrony varies greatly with latitude and correlates with EPF frequency in passerine species (Figure 4.2; Stutchbury and Morton 1995, Stutchbury 1998a). For temperate zone species synchrony is imposed by the short breeding season, resulting in a peak period when most females are simultaneously fertile. Average breeding synchrony for passerine populations in the temperate zone is 35%, with a range of 13–72% (Stutchbury 1998a). In many tropical species, females in the same population lay eggs weeks or months apart (Chapter 2) over a long breeding season, and as few as 8% of females in a population might be fertile at a given time (Table 4.1).

The exceptions prove the rule. Tropical species that do breed relatively synchronously, like the Clay-colored Robin, have abundant EPFs as we predicted they would (Stutchbury *et al.* 1998). The Blue-black Grassquit (*Volatinia jacarina*) is socially monogamous but males display conspicuously in clusters resembling a lek and likely obtain extra-pair matings (Almeida and Macedo 2001). The breeding season of this species is relatively short, presumably because it is granivorous and abundant seeds are not available until the end of the rainy season.

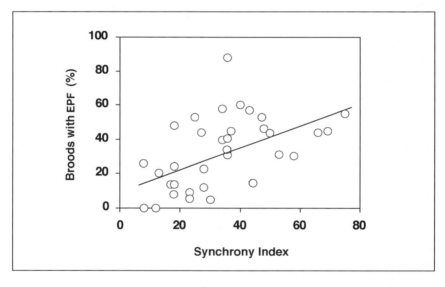

Figure 4.2
Relationship between frequency of EPFs and breeding synchrony in passerines (data from Stutchbury 1998a, and Table 4.1).

The key to understanding the evolution of extra-pair mating systems is female 'control' of EPFs. Females can benefit from EPCs and therefore seek out EPCs (review in Stutchbury and Neudorf 1998). When females seek EPCs, male and female interests within a pair conflict. Female control is well established in temperate zone birds (Hoi and Hoi-Leitner 1997, Neudorf *et al.* 1997) and has been shown clearly in a neotropical bird that exhibits female-defense polygyny, the Montezuma

Oropendola, *Psarocolius montezuma*. Females clump their pendulous nests in tightly packed colonies, presumably for protection against predators, as does the related and ecologically similar Yellow-rumped Cacique, *Cacicus cela* (Robinson 1986). Male oropendolas can then defend a group of females, and control copulatory access to them while they are at the colony (Webster 1995). Male size determines dominance, and the top male obtained over 90% of copulations at the colony. But despite their much smaller size, females have some control over fertilizations because they copulate with lower-ranking males away from the colony. The dominant male actually sires only one third of the offspring. Male Yellow-rumped Caciques use a different tactic to achieve social polygyny. They monopolize an individual fertile female and follow her everywhere (Webster and Robinson 1999). Breeding asynchrony among females allows the dominant male to be sequentially polygynous. Actual fertilization success of these dominant males and the extent to which females can obtain extra-pair copulations is unknown.

The breeding synchrony hypothesis argues that low breeding synchrony decreases the benefits to individual males and females of seeking EPFs (Stutchbury and Morton 1995, Stutchbury and Neudorf 1997, Stutchbury 1998a). There are many different kinds of benefits that females could gain from EPFs (fertility insurance, future mate assessment, good genes; Birkhead and Møller 1992). For females to benefit from EPFs by getting good genes, they must be able to judge the genetic quality of their own mate relative to other males. This is not an easy task under any circumstances, but is made easier by breeding synchrony. Females can judge the relative quality of males directly during male intrusions for EPCs and the ensuing male–male fights (Sheldon 1994, Neudorf *et al.* 1997, Stutchbury 1998b) and male pursuit chases of females (Hoi 1997). Females might use displays like song to compare males (Hasselquist *et al.* 1996). These are costly activities in terms of time or energy that are difficult for males to maintain throughout the breeding season, particularly when males must provide parental care. When female nesting is synchronized, females can more reliably compare neighboring males who are displaying and competing under similar conditions and constraints (Morton *et al.* 1998, Stutchbury 1998a). From the male perspective, when few females are fertile at one time (low synchrony) the high price of male–male competition and displaying outweighs the potential benefits to be gained from very few opportunities for extra-pair fertilizations.

A comparison of tropical and temperate zone birds will illustrate this nicely. Opportunities for mate assessment and gaining EPCs in the Hooded Warbler, *Wilsonia citrina*, which breeds synchronously, differs greatly from the tropical Dusky Antbird, *Cercomacra tyrannina*, which breeds asynchronously (Figure 4.3). In Hooded Warblers most females in the population lay their first clutch within a two week period. During the peak of female fertility, all the males in the population are competing simultaneously under similar constraints for a large number of potential fertilizations. Females can accurately assess male quality through their singing behavior and male–male battles during EPC attempts. Males seek EPCs, even when their own mates are fertile, because mate guarding is not effectual. Females seek EPCs off-territory where they cannot be guarded. In Dusky Antbirds, female fertility is not concentrated in time and males never have a large number of fertilizable females to compete for. At any point in time, some males are not nesting, others are incubating, and others are feeding young, making for a very uneven playing field which leaves it difficult for females to assess male quality.

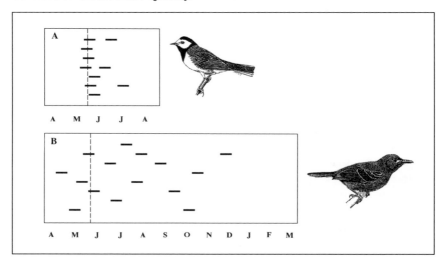

Figure 4.3
Temporal distribution of female fertile periods (solid bars) in A) a typical synchronously breeding temperate bird, the Hooded Warbler and B) a typical asynchronously breeding tropical bird, the Dusky Antbird. Dashed lines illustrate the difference in breeding synchrony. Drawings from Owings and Morton (1998) and Haverschmidt (1968).

Asynchronous breeding is a good explanation for the low frequency of EPFs in the tropics, but other ecological factors have also been suggested as important. Resident species might have lower EPFs than migrant species because the short time window for breeding combined with synchronized arrival results in rapid pair formation in migrants. Thus, some females unintentionally pair with low quality males, forcing them later to seek good genes from high quality males (Westneat et al. 1990, Slagsvold and Lifjeld 1997, Weatherhead and Yezerinac 1998, Petren et al. 1999). This idea seems reasonable, but it is flawed because it assumes that females of resident species have ample time for assessment of social mates and therefore all pair with high quality males (and do not need to seek EPFs). This is far from the truth. In socially monogamous species only one female pairs with each male, by definition. Even when mate assessment occurs year-round and pair formation can occur at any time, some females will be forced to pair with a male they know is low quality, or not breed at all. Longer period for mate assessment cannot explain why female tropical birds do not obtain EPFs. Temperate Black-capped Chickadees, *Poecile atricapillus*, pair months before nesting begins but still have EPFs (Otter et al. 1994).

Tropical passerines generally have larger territories and breed at lower density compared with ecologically similar temperate zone passerines (Terborgh et al. 1990). Thus lower density could be a factor in the lower extra-pair behavior of tropical birds, but only if density plays a big role in the costs and benefits of seeking extra-pair copulations. In a broad comparative analysis of birds, Westneat and Sherman (1997) found no correlation between EPF frequency and density. Studies within species often find no relationship between nearest neighbor distance and the occurrence of extra-pair young in the nest (Tarof et al. 1998). In some species there is a density effect, but the effect is not due to density per se. It is not that larger distances between territories make EPCs too costly to obtain but rather that high quality females that seek EPCs preferentially settle in high density situations (Hoi and Hoi-Leitner 1997). In many tropical passerines territories are not so far apart as to preclude movements between them for extra-pair matings. Even if territories are widely separated, extra-pair mating systems occur (Tarof et al. 1998). In the temperate zone Hooded Warbler, males sometimes breed in isolated forest fragments yet they fly across open agricultural fields to seek EPCs with females in other fragments up to 1.5 km distant (Norris and Stutchbury 2001).

Extra-pair behavior may have important conservation implications for tropical birds. In tropical forest birds there has been no history of selection for making long-distance forays because EPCs are not an important part of their mating system. Terrestrial insectivores are among the first birds to disappear from tropical forest fragments (Stratford and Stouffer 1999), in part because they are so unlikely to disperse across gaps and therefore only rarely colonize forest fragments.

The lack of extra-pair behavior in tropical birds is manifested by another major difference between latitudes. Not only do tropical passerines have smaller testes (Figure 4.1), they also differ dramatically from temperate zone birds in testosterone levels. The temperate zone pattern of high testosterone during territory establishment and pair formation, and low testosterone during offspring care, does not apply to tropical birds (Levin and Wingfield 1992, Wikelski *et al.* 1999b, Hau *et al.* 2000). Seasonal peak plasma T level in tropical birds (Figure 4.4; Wikelski *et al.* 1999a,b) ranges from only 0.3 ng ml^{-1} in Spotted Antbirds to 1.8 ng ml^{-1} in the Clay-colored Robin. Compare this with temperate zone passerines whose plasma T levels are in the neighborhood of 2.1 to 5.5 ng ml^{-1} (Wingfield 1984, Beletsky *et al.* 1989b, Ketterson *et al.* 1992, Johnsen 1998, Hau *et al.* 2000). Testosterone level is so low in breeding males of some tropical species that it is difficult to even detect (Wikelski *et al.* 1999a,b, 2000, Hau *et al.* 2000). Despite this, males vigorously sing, defend territories, attack intruders and pair with females, and females are also fiercely territorial and sing.

We view the high testosterone levels of temperate zone birds as an adaptation for competing for EPFs. Testosterone, a performance-enhancing steroid, may be necessary for males to maintain the exhausting battles and displays that are associated with strong sexual

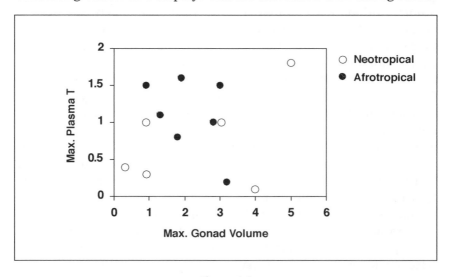

Figure 4.4
Peak plasma testosterone (T) levels (ng ml^{-1}) and relative gonad volume (mm^3 g^{-1}) of neotropical and afrotropical passerines (data from Wikelski *et al.* 1999b).

selection and mate choice common to extra-pair mating systems. Studies on temperate zone birds have shown that testosterone enhances extra-pair fertilization success (Raouf *et al.* 1997). High testosterone levels are very costly to maintain because testosterone apparently suppresses the immune system (Folstad and Karter 1992). But the benefits, in terms of extra-pair fertilizations, can be great. In contrast, the long, asynchronous breeding seasons makes EPF strategies unprofitable in the tropics. The absence of an extra-pair mating system in tropical birds makes high levels of testosterone unnecessary and useless.

Tropical birds vary greatly both in gonad size and in peak testosterone levels (Figure 4.4). It is too early to make sense of all the variations, but some consistent patterns emerge. Some permanently territorial and socially monogamous species, like Spotted Antbirds and Song Wrens, *Cyphorhinus phaeocephalus*, have tiny gonads and low testosterone. We predict that these species do not have an extra-pair mating system, and that is why they have no need for large gonads and high testosterone to enhance territory defense and mate attraction. Species with intense sexual selection, like the lekking Golden-collared Manakin, *Manacus vitellinus*, do have large gonads and high testosterone. The species with the highest plasma T level and relative gonad size (Figure 4.4) is the Clay-colored Robin. As we mentioned above, this is no coincidence. This robin is the only species in the sample that is known to have an extra-pair mating system (Stutchbury *et al.* 1998). Blue-gray Tanagers, *Thraupis episcopus*, are distinctive because they have large gonads but almost no testosterone! This species does not defend all-purpose territories, but the large gonad size might suggest intense sperm competition and sexual selection.

A comparison of two jay species supports the view that male–male competition for mating opportunities drives differences in testosterone between species (Vleck and Brown 1999). Western Scrub Jays, *Aphelocoma californica woodhouseii*, had a very short-lived peak in testosterone whereas the Mexican Jay, *A. ultramarina*, had a very prolonged period with elevated testosterone. The Western Scrub Jay is socially monogamous and studies on its Florida counterparts found genetic monogamy (Quinn *et al.* 1999), meaning relatively low male–male competition. Mexican Jays live in large social groups with multiple females. Extra-pair fertilizations are common throughout the breeding season (Brown *et al.* 1997), and male–male competition for matings is intense. This has resulted in prolonged elevated testosterone.

Schwagmeyer and Ketterson (1999) offered a different explanation for the connection between breeding synchrony and EPFs. They argued that testosterone mediates a tradeoff between male parental care and male EPC behavior. High testosterone suppresses parental behavior whereas low testosterone mediates male parental care (Ketterson and Nolan 1994). Low breeding synchrony means that when a particular female is fertile, many of her neighboring males would be caring for eggs or young and therefore be unavailable as extra-pair partners. In the Wattled Jacana, *Jacana jacana*, where females are larger than males and polyandrous, females do not obtain EPCs from males that are incubating (Emlen *et al.* 1998). Although male incubation may prevent males from seeking EPCs (Magrath and Elgar 1997), feeding young does not appear to constrain male EPF behavior. In some species males do feed young and seek EPCs at the same time (Pitcher and Stutchbury 2000) and EPF frequency does not correlate with male feeding effort across species (Schwagmeyer *et al.* 1999). If male care contrains EPF behavior, then we can ask whether high male parental care in tropical birds selects for low testosterone levels, which in turn results in low EPFs.

We believe that the cause and effect is different. The evolution of extra-pair mating systems is not favored under conditions of low breeding synchrony. Because of low levels of extra-pair behavior, selection does not favor high levels of testosterone. Without EPFs, selection favors high male parental care, and similar sex roles. Research on tropical birds should show which view is more efficacious.

4.2   Sex roles, male parental care, and sexual selection

Mating systems are intricately tied to sex roles in parental care and territory defense. Sex roles and sexual selection are tied in a feedback loop, reinforcing each other (Andersson 1994). When sex roles are highly divergent, selection pressures are divergent and one often sees sexually selected traits and behaviors in males but not females. Sex role divergence is typical of temperate zone passerines, where males sing and defend territories, compete for EPFs, and their parental care is generally limited to nest defense and feeding young. Females, of course, do not usually sing or defend territories as aggressively as males do and they build nests and incubate the eggs alone. Extra-pair behavior creates strongly divergent sex roles as males compete for EPFs, females are discerning in their acceptance of extra-pair mates, and males and females have an unequal genetic stake in the brood they raise together.

There are several models for the theoretical relationship between extra-pair paternity and male parental care (reviewed in Wright 1998). The optimum level of male parental effort is reduced, when paternity in the brood is low, because males achieve a greater overall fitness by investing their time and energy in somatic effort (living longer) or mating effort (seeking extra-pair matings). The exact nature of the relationship depends on how crucial male care is for offspring survival and recruitment (Whittingham *et al.* 1992b) and how extensively male parental effort trades off with both male EPC effort and future male survival (Westneat and Sherman 1993). Little is known about life history tradeoffs or the importance of male care in tropical birds (Chapter 3). But, all the models predict that a high frequency of EPFs will select for reduced male parental care, which in turn will reinforce sex role divergence.

Sex roles are diverse and variable in tropical birds but are more often similar than is the case with temperate passerines (e.g., Greenberg and Gradwohl 1983). In many species, like the Dusky Antbird, females and males have almost identical roles in territory defense and parental care. Even though males cannot lay eggs, they might make up for this gender disparity by extensive courtship feeding (Skutch 1996). Dusky Antbirds exhibit convergent sex roles because they do not have extra-pair behavior (Fleischer *et al.* 1997). We believe the lack of extra-pair behavior in tropical birds underlies why males often build nests, incubate eggs, and brood young while female territory defense is as vigorous as male defense (Chapter 5) and females often sing (Chapter 6).

Sex role convergence is also seen clearly in tropical ducks. Temperate zone migratory ducks are characterized by frequent forced EPCs, relatively synchronous and short breeding seasons, seasonal pair bonds and no male parental care (McKinney *et al.* 1983, McKinney 1985). Each year new pair bonds form on the wintering grounds and, once the pair arrives in the breeding area, males guard their mates closely until laying is complete. Then they abandon their mate. Forced EPCs are frequent and often violent. The mating system and parental care system of tropical and southern hemisphere ducks differs markedly. Pair bonds often persist year round, breeding seasons are prolonged and males of many species provide parental care (McKinney 1985, Williams and McKinney 1996, Johnson *et al.* 1998). We expect that EPFs would be uncommon in these species, and that sex role convergence results from the absence of extra-pair behavior. The White-cheeked Pintail, *Anas bahamensis*, on the Bahama Islands illustrates the importance of extra-pair behavior, as opposed to tropical latitude by itself, in the evolution of gender roles. It is sedentary and exhibits year-round courtship, but no male parental care (Sorenson 1992). Competition for females is intense owing to a skewed sex ratio, and forced extra-pair copulations are common.

What does high male parental care and sex role convergence mean for mate choice and sexual selection in monogamous passerines? It would be a mistake to conclude that the absence of EPFs equates with an absence of sexual selection in monogamous birds (Cunningham and Birkhead 1998). Instead, both males and females experience similar sexual selection pressures because they have similar sex roles (Andersson 1994). When parental care by both sexes is equal, both genders are expected to be choosy in selecting mates and to compete intra-sexually for high quality partners (Trivers 1972, Jones and Hunter 1993).

Sexual selection in monogamous birds can also arise when individuals, because of the high quality of their mates, increase their parental effort (Burley 1988). This 'differential allocation', is thought to be an adaptation by females that depends on mate quality. When gender roles are similar, 'differential allocation' should characterize both female and male parental effort equally. Differences among pairs in timing of breeding, clutch size, or relative parental effort may be linked to mate quality (reviewed in Cunningham and Birkhead 1998). Individuals may also be able to increase their chances of attracting a mate by offering a higher level of parental care. For instance, temperate male Blue-headed Vireos, *Vireo solitarius*, build nests as part of their courting behavior (Morton *et al.* 1998). In the tropical Buff-breasted Wren, *Thryothorus leucotis*, males invest heavily in nest building, particularly for dormitory nests (Gill and Stutchbury, unpub.). Male nest building may be a sexually selected trait, because male nest building effort indicates a male's future feeding effort. If so, one would predict that females would be more likely to divorce males with low nest building effort.

Little is known about mate choice and parental care in monogamous tropical birds. Most passerines are paired year round, so mate assessment is continual and mate switching can occur any time. But what cues do individuals use to assess mate quality? Song, plumage coloration, displays or some other trait? Is parental effort an important trait used in mate choice? Given the very high rate of nest predation in the tropics, parental nest defense or ability to renest quickly, rather than feeding or incubation effort, may be more important to nesting success. Low food availability and prolonged period of fledgling care in many tropical species suggests parental effort to fledglings may be much more important than care at the nest, as has been shown for some temperate zone species (Wolf *et al.* 1988, Evans Ogden and Stutchbury 1997). Empirical data and field experiments are sorely needed.

Another view is that the importance of mate choice is rather overblown as a result of the preponderance of breeding territoriality in temperate birds (Chapter 5). Mate choice might be coincidental to territory selection. For example, Dusky Antbirds leave current territories and mates to move to territories more apt to enhance individual longevity (Morton *et al.* 2000). The new territory is always already occupied by a mate, which pairs with the newcomer automatically. There may be a high correlation between the quality of a bird as a mate and the quality of the year-long territory it occupies. These switches do not appear to be related to the mate on the new territory or to increased nesting success.

4.3 Contrasting the mating system and behavior of two *Elaenia* flycatchers that differ in breeding synchrony

A good way to illustrate how synchrony, EPFs, and mating behavior are closely tied is to compare two closely related species that breed in the same habitat but differ in breeding synchrony, as Morton *et al.* (1998) did for temperate zone vireos. The Yellow-bellied Elaenia, *Elaenia flavogaster*, is a typical tropical bird in having year-round pair bonds and territory defense. Females sing and duet with the male, males assist with nest building and feeding young, but not with incubation, and breeding is asynchronous. The Lesser Elaenia, *E. chiriquensis*, is anything but lesser! Skutch (1960) calls this the Bellicose Elaenia for good reason, because chases and fights are commonly seen. In contrast to the Yellow-bellied, this elaenia is an intratropical migrant, breeds relatively synchronously, females do not sing, and males feed the young but do not build nests or incubate. We predict that the Lesser Elaenia has an extra-pair mating system, due to higher nesting synchrony, while the Yellow-bellied Elaenia is genetically monogamous. In addition to being close relatives, both species nest in the same habitat and at the same time during the dry season, and both feed on fruit as adults.

Our study in Panama (Morton *et al.* unpubl) found that Lesser Elaenias have a breeding synchrony index almost twice as high as Yellow-bellied Elaenias (15–18% vs. 9%) (Table 4.2). The Lesser Elaenias are similar to temperate zone 'EPF species' in that they have larger testes, and we will soon know if they also have more EPFs.

Table 4.2

Comparison of the breeding synchrony, testes size and breeding behavior of the congeneric Lesser Elaenia (*Elaenia chiriquensis*) and Yellow-bellied Elaenia (*E. flavogaster*).

	Lesser Elaenia	Yellow-bellied Elaenia
Breeding Synchrony (%)[a]	15–18%	9–10%
EPFs (% young)	?	?
Testes Mass/Body Mass	0.02	0.007
Male Intrusion Rate (#/h)	0.3	0
Male Pairing Success	80%	100%
Mate Following by Male[b]	2%	7%
Male Nest Defense[c]	4%	30%
Male Nest Building[d]	0%	17%

a: percentage of females simultaneously fertile
b: percentages of trips to or from the nest that male followed the female closely
c: percentage of nest defense events where the male participated
d: percentage of nest building trips performed by the male

Male pairing success was only 80% (9 of 43 males never got mates) compared with 100% pairing success for Yellow-bellied Elaenias. Unmated male Lesser Elaenias had a remarkably high song rate (> 150 songs h^{-1}) compared with paired males (typically < 10 songs h^{-1}). Male Lesser Elaenias frequently intruded onto neighboring territories where females were fertile (e.g. Stutchbury 1998b), but in Yellow-bellied Elaenias we never saw single males intruding onto territories. Instead, territorial disputes occurred at boundaries and involved each pair duetting and chasing the other pair. Mate guarding did not occur in either species. This example illustrates that ecological differences between closely related species, occupying the same habitat, can result in predictable differences in an entire suite of traits that we predict coevolve as part of an extra-pair mating system.

4.4 Promiscuity

Although social monogamy predominates in tropical and temperate birds, tropical birds are well known for their promiscuous mating systems. Spectacular leks and mating systems occur most notably in the hummingbirds, manakins, cotingas, birds of paradise and bowerbirds. Understanding how and why promiscuity evolves from monogamy has been a central theme in behavioral ecology. Mating systems can be viewed as a continuum of male parental effort, ranging from monogamous species where males care for the young, to polygynous species where males defend territories but care is minimal or non-existent, to lekking species where females associate with males only for copulations. Pioneering work by Crook (1964) on weaverbirds (Ploceinae), an Old World tropical group, first established a link between diet and mating system. Insectivorous forest weaverbirds are usually monogamous, whereas gramnivorous savannah weaverbirds are usually polygynous. Food is presumably more limiting for insectivorous birds which favors males to pair monogamously and assist with feeding the altricial young.

Why do the tropics have so many promiscuous species? The tropical passerines best known for promiscuous mating systems are frugivorous. Lack (1968) and others (Lill 1974, Snow 1976b) suggested that polygamy and lekking has evolved more often in the tropics because males are 'emancipated' from the constraints of providing parental care owing to the abundance of fruit. Fruit is easier to locate and harvest than arthropod prey because fruit 'wants to be eaten' (Chapter 7). Females of frugivorous species should be able to raise young alone more easily than can females of insectivorous species. Leks should be more common than resource defense polygyny (Emlen and Oring 1977) because males rarely can control access to abundant and widely dispersed fruit. Other aspects of frugivory may also increase the potential for promiscuity to evolve. Frugivory may allow males to spend a large fraction of their time seeking and attracting mates, rather than searching for food (Snow 1971). Frugivory is also associated with large home ranges of females and males, which may increase the rate at which males can encounter and successfully court females (Bradbury 1981).

The idea that frugivory, and plant diets in general, predisposes a species to become promiscuous can be tested by comparing the diets of close relatives that differ in mating system. Monogamous species are expected to be insectivorous (i.e., their diet requires male help for feeding young) whereas promiscuous species should have a diet (fruit, nectar) that allows females to raise young alone.

Among the New World tropical flycatchers (Tyrannidae) monogamy is the rule even though many species are highly frugivorous and feed a large amount of fruit to their young (Skutch 1960). The Ochre-bellied Flycatcher, *Mionectes oleagineus*, and several of its congeners, is peculiar because males display solitarily or in small leks to attract females

(Westcott 1997). This species likely eats more fruit than other forest-dwelling flycatchers but there are also insectivorous forest flycatchers that apparently do not form pair bonds or have male parental care (Snow and Snow 1979). In particular, Skutch (1960) suggested the Northern Bentbill, *Oncostoma cinereigulare*, has a promiscuous mating system despite being entirely insectivorous.

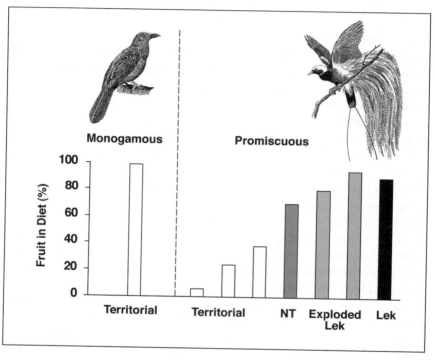

Figure 4.5
Percentage of fruit in diet for birds of paradise that are monogamous and terri-torial versus polygynous and territorial, non-territorial (NT), dispersed in an exploded lek, or clumped in a lek. Data from Beehler (1985) and Beehler and Pruett-Jones (1983). Drawings from Lack (1968).

The birds of paradise provide another interesting example (Figure 4.5). As expected, species with exploded or communal leks all have a large fraction of fruit in their diet (Beehler and Pruett-Jones 1983). However, the promiscuous Black-billed Sicklebill, *Epimachus albertisi*, where males display solitarily, has an almost entirely arthropod diet (Beehler and Pruett-Jones 1983). The Trumpet Manucode, *Manucodia keraudrenii*, is monogamous and has male parental care. Rather than being insectivorous, the Trumpet Manucode is strictly frugivorous even when feeding young (Beehler 1985). In contrast to the other birds of paradise, this species specializes on fig fruit which has low nutritional value, is relatively uncommon and widely dispersed within the bird's territory (Beehler 1983). The young may require such a large quantity of fruit to develop at a normal rate that both parents are needed (Beehler 1985). Although most birds of paradise are frugivorous and promiscuous, other frugivorous taxa in New Guinea (cuckoo-shrikes, honeyeaters, berrypickers) are monogamous (Beehler 1983).

In the Eurylaimidae family endemic to Madagascar, the Velvet Asite, *Philepitta castanea*, has a dispersed lek mating system and eats fruit as well as nectar (Prum and Razafindratsita 1997). Although this is consistent with the fruit-lekking link, two other species in this family form pair bonds and eat nectar. Why fruit, and not nectar, would be more likely to promote lekking is not clear. Hummingbirds as a group are promiscuous and many tropical species have leks (Snow 1974).

These examples show that the connection between diet and promiscuity is not as strong as earlier literature suggested. This may be due to the fact that, no matter how frugivorous adults are, most still feed insects to their nestlings (Chapter 7). The key to mating system evolution is not only the diet of adults, but whether females can successfully breed without male parental care. An important experimental tool to determine how important male parental care is to reproductive success is male removal. Most removal studies in temperate monogamous species have found that male parental care does increase reproductive success (Wolf *et al.* 1988). Such experiments have not been done with tropical species. Do females of insectivorous flycatchers have more difficulty raising young alone than females of frugivorous flycatchers? Are male Trumpet Manucodes really needed to raise a brood successfully on fig fruit alone?

If fruit is so easy to harvest and can be fed to nestlings, why are so many frugivorous species monogamous? Even in frugivorous birds, nestlings are often fed arthropods early in their development (Morton 1973). Many frugivores breed during the dry season when arthropod abundance is low, and male parental care may be crucial only for a few days when nestlings cannot survive on fruit alone. This could be tested with temporary male removal experiments in a species like the Clay-colored Robin where starvation in nestlings is common even when both parents are feeding (Chapter 3).

4.5   Cooperative breeding

Cooperative breeding, where many individuals can assist a single pair in breeding or breed together themselves, lies at the other extreme of sociality from promiscuous species. Cooperative breeding is rare in north temperate land masses, but is much more common in the tropics worldwide and Australia (Brown 1987). The first description of helpers at the nest came from none other than Skutch (1935)! Cooperative breeders are extremely diverse in terms of taxonomy, habitat, and the details of the mating system (who helps, how many breeders, etc.), so searching for underlying ecological determinants of helping behavior is a challenge. Long-term studies of cooperative breeding have been done on tropical birds as diverse as the Pied Kingfisher, *Ceryle rudis* (Reyer 1990), Green Woodhoopoe, *Phoeniculus purpureus* (Ligon and Ligon 1990), White-fronted Bee-eater, *Merops bullockoides* (Emlen 1981), Galapagos Hawk, *Buteo galapogoensis* (Faaborg 1986), Groove-billed Ani, *Crotophaga sulcirostris* (Vehrencamp 1978), and Stripe-backed Wren, *Campylorhynchus nuchalis* (Rabenold 1990).

The evolution of cooperative breeding can be understood by examining two distinct questions (e.g. Brown 1987). First, why do sexually mature individuals choose to delay breeding and stay home? Second, why do these individuals then help to raise the young on that territory?

Why stay at home?

As with many other topics discussed in this book, the question can be turned on its head to ask why cooperative breeding is rare in the temperate zone, rather than why it is common in the tropics. The key feature of tropical birds that promotes cooperative breeding is permanent residency which allows young birds the option of living with their parents, or other adults (Brown 1987). Many temperate zone birds are long distance migrants, which forces young birds to disperse. Individuals cannot stay at home if there is no 'home'. There are year-round residents in the temperate zone, and it is more interesting to consider why cooperative breeding is not more common among these species.

In resident species, young birds do have the option of staying home, but under what conditions would this be favored over dispersing and breeding immediately? High adult longevity is typical of many tropical birds (Chapter 3) and promotes delayed dispersal in two ways. First, delayed breeding is a strategy of the future and is more likely to eventually pay off for long-lived birds (e.g. Wittenberger 1979). Second, high adult survival results in a low turnover on territories and few breeding vacancies that can be filled by young birds (Brown 1987). When the production of young exceeds the number of breeding vacancies, successful dispersal becomes difficult for most juveniles owing to habitat saturation.

The habitat saturation model for the evolution of cooperative breeding is based on the idea that young birds delay dispersal because, in its strictest form, no breeding habitat is available (Emlen 1982, Brown 1987). There are several reasons why young may stay home, discussed by Stacey and Ligon (1991) as 'benefits of philopatry.' Young may opt to stay home if only marginal or low quality territories are available, which can be considered another form of habitat saturation because high quality territories are limiting. In either case, staying home is more advantageous than dispersing and breeding elsewhere. This can occur when any critical resource is scarce, including roosting sites and mates. What evidence is there that habitat saturation actually occurs in tropical cooperative breeders?

Many studies have demonstrated variation in reproductive success among groups, and correlated this with some feature of the habitat (e.g. Langen and Vehrencamp 1998). While this implies a likely limitation of quality territories, it alone is not strong evidence for the habitat saturation model. A stronger approach is to compare populations or species that differ in mating system, and predict that the cooperatively breeding populations will experience greater habitat saturation. Zack and Ligon (1985a,b) compared two species of congeneric shrikes (*Lanius* sp.) in Kenya, and found that the cooperative species occupied dense woodland, a limited habitat with high food abundance during the dry season that results in high adult survival. These conditions result in a short supply of high quality territories and few good options for dispersal for young birds. The non-cooperative species occupies ubiquitous open habitat with lower food availability, low adult survival, and abundant breeding territories for young birds. In a unique translocation experiment (Komdeur 1992, Komdeur *et al.* 1995) found that Seychelles Warblers, *Acrocephalus sechellensis*, transferred to unoccupied islands showed no cooperative breeding until the high quality territories were filled. Individuals born on high quality territories were unlikely to disperse to lower quality territories and instead remained home to help, but young born on poor territories dispersed to breed on poor territories.

An experimental test of the 'benefits of philopatry' idea involves the removal of breeders in territories of differing quality to test whether vacancies on high quality territories are more likely to get filled quickly.

Zack and Rabenold (1989) found that in Stripe-backed Wrens, *Campy-lorhynchus nuchalis*, more females fought, and fought more vigorously, for experimentally created breeding positions in high quality territories than in low quality territories. In this species territory quality is determined by group size (number of helpers) rather than an intrinsic feature of the territory itself.

Resources other than food, such as nest sites and roosting sites, can also result in limited opportunities for successful dispersal (Restrepo and Mandragón 1998). For instance, cooperative breeding is relatively uncommon among fruit specialists, perhaps because of the loose territorial system seen in many frugivores owing to the abundance and changing availability of fruit and the difficulty of defending it (Brown 1987). Cooperative breeding does occur in frugivores like hornbills (Witmer 1993), and toucan barbets (Restrepo and Mandragón 1998), which defend nesting and roosting cavities.

A limitation of quality territories is not always the answer to delayed breeding (Macedo and Bianchi 1997). Ecological factors that favor group living are not restricted to habitat saturation (Koenig and Pitelka 1981), and include group defense of nests (e.g. Cuckoos), group foraging in anis and cuckoos (Vehrencamp 1978, Macedo and Bianchi 1997), short and unpredictable food supply for raising young so that helpers are essential (e.g. Bee-eaters), or scattered and clumped food where group defense is advantageous (e.g. Acorn Woodpeckers, *Melanerpes formicivorus*). Nevertheless, the majority of studies on tropical cooperative breeders have found evidence of a limitation of quality territories or mates, which favors delayed dispersal by young (e.g. Emlen 1981, Zack and Rabenold 1989, Ligon and Ligon 1990, Curry and Grant 1990, Reyer 1990, Strahl and Schmitz 1990).

Why help?

For helping behavior to evolve, young birds must gain some direct or indirect benefit from helping. In many cooperatively breeding species the helpers are prior offspring of the breeding pair (reviewed in Cockburn 1998), suggesting that kin selection can be an important benefit of helping. But in many species helping by young does not result in an increased production of nondescendent kin (Cockburn 1998) and, instead, helpers may benefit directly. In many species helpers end up breeding on their natal territory, or an adjacent one, indicating that staying at home is a direct route to breeding independently.

The ecological conditions that favor genetic monogamy in tropical birds help to set the stage for cooperative breeding to evolve. Indirect benefits to helpers are only possible if they are related to the young they help. While extra-pair paternity affects only who fathers the young on a territory, even modest levels of EPFs significantly reduces the average degree of relatedness between helpers and the young they assist. Most studies of tropical cooperative breeders have found that extra-group paternity is rare, generally less than 5% of young (Rabenold *et al.* 1990, Haydock *et al.* 1996, Cockburn 1998, Conrad *et al.* 1998). In some species, male helpers gain EPFs with the breeding female (within-group EPFs) but in this case some of the offspring they help are actually their own (Rabenold *et al.* 1990) so kin selection does not apply. The high levels of EPFs found in most temperate species, even year-round residents, would be an impediment to the evolution of cooperative breeding.

What kind of help do helpers provide? For many tropical species, food availability for feeding young appears to be limiting (Chapter 3) and helpers increase food delivery rates to the nest (e.g. Emlen 1981). In Pied Kingfishers, *Ceryle rudis*, (Reyer 1990) unrelated helpers are tolerated by breeding pairs only in populations with low food supply or when brood size has been experimentally increased, indicating that helpers are needed to raise young. Tropical birds also experience high rates of nest predation (Chapter 3), and helpers in many cooperative species play a key role in nest defense (Austad and Rabenold 1985, Innes and Johnston 1996, Restrepo and Mandragón 1998).

Why isn't cooperative breeding more common in the tropics?

How well does habitat saturation explain the evolution of cooperative breeding in the tropics? It is clear that most cooperatively breeding species experience some form of ecological constraint that favors delayed dispersal. But, many tropical species that do not breed cooperatively nevertheless have the key ingredients that promote cooperative breeding. These include long lifespan, year-round territoriality, and variation in territory quality. So why isn't cooperative breeding more common?

One answer is that the variation in territory quality is not extreme enough to favor delayed dispersal. This is a sort of 'cooperative breeding threshold model', akin to the polygyny threshold model that explains why females should settle polygynously on high quality territories rather than pair with a monogamous male on a poor quality territory (Verner and Willson 1966, Orians 1969). Breeding success on a poor territory must be dismal before kin selection benefits or direct benefits through inheritance of a good territory can offset the costs of not breeding at all. Many of the cooperative breeders studied to date occupy open habitats, which facilitates observation (Brown 1987). But this bias may also mean that territories vary more in quality, especially for food abundance during the dry season, than might occur in forest habitats. Forest birds may not experience sufficient extremes of territory quality to favor delayed dispersal. This may also apply to resident temperate zone species, where food abundance during the breeding season is high for most pairs and raising young is possible on all territories.

For most tropical birds, those that do not breed cooperatively, little is known about the variability in territory quality or how (and if) territory quality affects adult survival and reproductive success. Furthermore, little is known about the dispersal strategies of young birds (see Chapter 5). The presence of 'floaters', where young delay breeding, is good evidence for a limitation of quality territories. In Rufous-collared Sparrows, *Zonotrichia capensis*, young birds do delay breeding but do not help, and instead wait furtively on territories for vacancies to occur (Smith 1978). In Dusky Antbirds, removal experiments indicated floaters were rare because experimental vacancies often went unfilled or were filled by neighboring territory owners rather than previously non-territorial birds (Morton *et al.* 2000). Comparisons of close relatives that differ in cooperative breeding, like Zack and Rabenold's study on *Lanius* shrikes in Kenya, are the most promising way to understand how ecological differences lead to different mating systems.

5 | Territoriality

Latitudinal differences in territorial behavior in relation to breeding seasons and mating system have had a great influence on latitudinal differences in life history strategies. In the temperate zone, territory establishment by the male, followed by mate attraction is the common model of territory defense (Freed 1987). Temporary breeding territories, coupled with high over-winter mortality, mean that males and females have many unoccupied areas for establishing territories when they return in spring. For those that do return about 50% of their neighbors will be altogether new to them. Males and females have very divergent interests due to the prevalence of extra-pair matings (Chapter 4). In many species one or both sexes sneak off their territory to visit neighbors for extra-pair matings (Neudorf *et al.* 1997, Stutchbury 1998b). Females seek out extra-pair matings, whereas males try to prevent their mates from doing this while at the same time seeking EPCs themselves. Males must defend their territories from frequent and sexually-motivated intrusions, even after territory boundaries are well established. Territorial aggression is highest early in the nesting season when territories are being established and EPC competition is at its peak (Arcese 1987).

In the tropics year-long territory defense is common and adult survival high, so breeding vacancies may be scarce. Consequently, birds face very different constraints in choosing mates and territories. It is common that either gender can maintain a territory, if their mate dies or deserts them, and attract a new mate (e.g. Morton *et al.* 2000). Males and females each have similar interests in defending their territory against same-sexed neighbors, who are after real estate rather than sex. Actual territorial intrusions are relatively infrequent, but border disputes and territory defense by singing can persist throughout the year. Little is known about how territorial aggression varies over the season, for either gender. Tropical territorial systems are not well studied, and are much more diverse than the 'simple' year-round defense described above. The costs and benefits of territory defense depend on the type of territory and what is at stake. The general temperate zone model, which is constrained by climate and so driven by extra-pair mating behavior, applies to only a minority of the tropical species.

5.1   Territory systems

Over 90% of North American passerines have a similar territorial system, they defend breeding territories for only a few months each summer (Table 5.1). In the tropics, as usual, diversity is the name of the game. Only 13% of passerines defend territories only for the breeding season. Instead, the predominant territorial system is year-round defense of feeding and nesting territories (Table 5.1). This territorial system occurs even in more seasonal habitats such as mangroves (Lefebvre *et al.* 1992). The types of resources defended, and when they are defended, is highly variable in the tropics owing to year-round food availability and the defensibility, or lack thereof, of different food types. Generally, insectivores defend year-round all-purpose territories, while frugivores do not (Morton 1973, Buskirk 1976). Year-round territoriality and pair bonding is typical of tropical insectivorous birds, and occurs in 63% of passerines in Panama (Table. 5.1) and 40% in tropical South Africa (Rowan 1966). Arthropod resources are defensible because they are more or less evenly distributed spatially and temporally.

Table 5.1

Territorial systems of Panamanian passerines compared to North American passerine birds.

Type of Territory[a]	Number of species		Number of genera[b]	
	Panama	NA	Panama	NA
Breeding	42	224	28	89
Year-long	142	15	84	13
Army ant influenced	11	0	9	0
Mixed species flock	65	0	40	0
Fruit influenced	43	0	20	0
Lek	28	0	19	0
Total	**331**	**239**	**200**	**102**

a: See text
b: Species in some genera fit more than one territory type (e.g., *Elaenia*, *Vireo*, *Basileuterus*, *Sporophila*).

But there are many variations within this basic pattern. In addition to year-long territorial and permanent pair bond systems, the tropics offer:

1. many types of lek and cooperative breeding examples,
2. loosely defended territories where pair members leave well-defended nesting territories, more or less independently, to visit fruit sources,
3. species more or less dependent upon the peregrinations of army ants, and
4. the pros and cons of membership in mixed flocks of several different sorts.

In mixed flocks of the forest interior insectivorous type, many species are represented by only a pair or small family party. Tanager/flycatcher flocks of the forest canopy are more frugivorous but might still be represented by small numbers of each species. In contrast, fruit-eating birds of more open country, or llanos, characteristically consist of large groups of conspecifics (Moynihan 1962, Buskirk 1976, Morton 1979a, Munn and Terborgh 1979, Powell 1979). These are discussed further in Chapter 7.

The amazing diversity of territory types in tropical forests is best seen by comparing species within the same forest (Figure 5.1). Species that live in canopy flocks, like the White-shouldered Tanager, *Tachyphonus luctuosus*, have territories about four times the size of species that live in understory flocks, like the White-flanked Antwren, *Myrmotherula axillaris* (Figure 5.1A, B). Obligate ant-following birds like the Sooty Antbird, *Myrmeciza fortis*, have huge territories (Figure 5.1D) compared with other insectivorous birds that defend year-round territories as pairs or family groups (Figure 5.1C). Lekking species are common, and either form traditional leks where many males display very close together (e.g. many hummingbirds and manakins), or they court as individuals on solitary display perches (Figure 5.1E) as in the Dwarf Tyrant-Manakin, *Tyranneutes stolzmanni*.

Territory sizes for tropical birds are typically larger than what we see for comparable birds in the temperate zone (Terborgh *et al.* 1990). For instance, forest flycatchers, wrens, robins, tanagers and vireos have territories that are typically 5–15 ha in size. This is generally 10 times larger than the area defended during the breeding season by their temperate zone counterparts (Terborgh *et al.* 1990). Males and females each defend the territory against same-sexed challengers, and do not

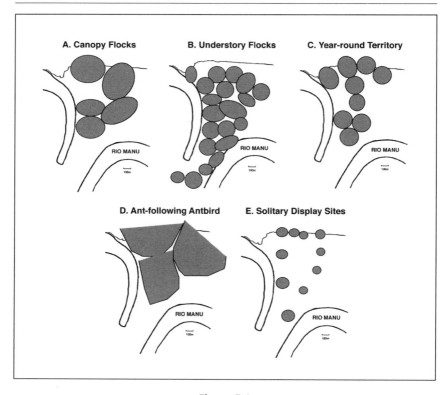

Figure 5.1
Representative territory types in a floodplain forest in Amazonian Peru
(Terborgh *et al.* 1990). Shaded areas are territories.

usually cooperate in defending the territory (Greenberg and Gradwohl
1983, Freed 1987, Morton and Derrickson 1996).

Turnover of adults on these territories is very low and this contributes to remarkably stable territory boundaries (Greenberg and
Gradwohl 1997). Lefebvre *et al.* (1992) suggested that territory stability may be low in seasonally versus permanently territorial tropical
species. However, we have found high site fidelity for several species
that seasonally defend nest site territories. Clay-colored Robins, *Turdus
grayi*, defend territories only during the dry season, and males and
females bred on the same (33/36) or adjacent territory (3/36) in subse-
quent breeding seasons (Stutchbury *et al.* 1998). Lesser Elaenias,
Elaenia chiriquensis, intratropical migrants that breed in the dry season,
also have high site tenacity with most returning adults renesting on the
same (6/9) or an adjacent (2/9) territory (Morton *et al.* unpubl). Thus,

territories are stable from the perspective of the individual that reuses its territory even in species with breeding territoriality. Another contributor to stable neighborhoods is that tropical species with year-long territories do not expand territorial boundaries, even when given the opportunity to do so experimentally (Morton *et al.* 2000). The reason for this is unstudied, but perhaps birds that are familiar with their territories are better able to avoid predators (Lima 1998).

Factors that determine territory quality for tropical birds are little studied. Food availability during the nonbreeding portion of the year may be more important in determining territory quality and size than food during the breeding period. The reason is that breeding success is often low. A bird might fledge young only once in its life. We speculate that territory quality will be based upon nonbreeding factors when annual reproductive success is less than 10%. Then, territorial quality that increases individual survivorship during periods of low food abundance, often the dry season, will prevail (Morton *et al.* 2000). Territory switching in Dusky Antbirds, *Cercomacra tyrannina*, was related to increasing adult lifespan and not to reproductive success per se (Morton *et al.* 2000).

Interspecific territoriality sometimes occurs where closely related species (congeners) compete for year-long territories in the most productive habitats. Robinson and Terborgh (1995) documented interspecific territoriality using reciprocal heterospecific playbacks in 10 of 12 species of non-oscine passerines in Peru. Most of these had non-overlapping territories. Some of the same species do not have interspecific territoriality elsewhere (Stouffer 1997).

Many insectivorous species that are permanently territorial feed in mixed-species flocks (Powell 1985). It is the foliage-gleaning and bark-gleaning birds that closely scrutinize substrates that are most tied to mixed-species flocks, because predator vigilance is difficult to maintain with this type of foraging (Powell 1985, Thiollay 1999). Mixed-species flocks allow birds to feed efficiently while taking advantage of the vigilance of the flock (Willis 1972). Generally a flock contains only a single family group of a particular species, due to strong territoriality. Some species defend permanent territories smaller in size than the flock, and the local territory owners join and leave the flock as it moves across territory boundaries. This means that a given territorial pair must spend much time foraging alone on its territory, while the flock is elsewhere. Other species have territories that conform to the territory boundary of the multi-species flock (Munn and Terborgh 1979, Power 1979, Gradwohl and Greenberg 1980). Some species (e.g. antwrens) sometimes defend seemingly excessive territories, and have very low population density. This is because the flock's boundaries are determined by the flock's larger species. Individuals of some smaller species benefit greatly by defending the entire home range of the flock so that they can join it at any time.

Other permanently territorial insectivorous passerines are professional ant-followers (Willis 1967, 1972). The sheer abundance and spatial concentration of food at ant swarms makes it uneconomical to exclude conspecifics. Although multiple pairs of a given species may be present at an ant swarm, the male and female owners of the territory where the ants are passing through are socially dominant over conspecifics. As the ants move into a neighboring territory, the owners of that territory become socially dominant at the swarm.

Many of the 42 species of Panamanian passerines exhibiting breeding territoriality (Table 5.1) are frugivorous birds, which defend small nesting territories but feed off-territory on fruiting trees. However, 13% of the species in Panama are permanently paired and defend year-long territories but leave them to visit fruit sources. In Table 5.1 this type of territoriality is called 'Fruit Influenced' to emphasize that fruit sources are ephemeral and not defended, even though the bird species have year-long territories and pair bonds (e.g. Yellow-bellied Elaenia, *Elaenia flavogaster*) or year-long pair bonds without year-long territories (e.g. Blue-gray Tanager, *Thraupis episcopus*). This contrasts with army ant influenced territories where pairs also leave their territories to feed at swarms, because the food supply (fruit) is not defended by anyone even if it happens to occur within a pair's territory. For instance, Clay-colored Robin males will not attack other robins that enter their territory to feed on fruit, but only if the visitors do not sing while they are there! Many other frugivores lek and do not even defend nesting territories.

Nectar is defensible because, unlike fruit, the food supply is rapidly renewed in flowers. In hummingbirds, long-term feeding territories are usually defended by males only (Wolf 1969, 1975). Intruders, male and female alike, are chased and attacked vigorously. In several species males allow females to enter the territory to feed, but only if the females allows the male to court her. In some species flowers are too scattered to be defended, so males 'trap-line' by defending a series of high quality flowers that they revisit on a predictable route.

Why aren't female hummingbirds also territorial? The spatial concentration of flowers allows males to defend nectar not only from females, which are smaller, but also from other species of hummingbirds. This interspecific territoriality means that males can monopolize high quality flower clusters. Females are forced to feed from scattered flowers that are not economically defensible. The few hummingbirds where females defend long-term feeding territories feature bright female coloration (for defense) and are either large so females can dominate smaller hummingbird species or occur on islands where there are few interspecific competitors (Wolf 1969, 1975).

Territory defense of food by both sexes, and competition between the sexes for limited food resources, can drive the evolution of sexual differences in resource use. The Purple-throated Carib (*Eulampis jugularis*), a hummingbird, is the sole pollinator of two *Heliconia* species (Temeles *et al.* 2000). Remarkably, specialization by each sex of the hummingbird on different species of *Heliconia* has caused sexual dimorphism in bill size and shape.

5.2   Territory defense

Defense and extra-pair copulations

Territory defense by temperate zone passerines is strongly influenced by the high frequency of extra-pair matings. In spring, males and females must acquire or reclaim a breeding territory when they first arrive. Before the discovery of extra-pair mating systems, much of territorial behavior was thought to revolve around establishing and maintaining the boundaries so neighbors did not take over part of the territory. Once boundaries are set after a few weeks, strangers (e.g. floaters) were thought to be a bigger threat to the territory than neighbors (e.g. Wiley 1991). We have a different view of territoriality now. Males and females make frequent and covert forays onto neighboring territories, not to take over the territory, but to seek copulations. In Hooded Warblers, *Wilsonia citrina*, males and females leave their territory about once every two hours to sneak onto a neighbor's territory (Neudorf *et al.* 1998, Stutchbury 1998b). These are high stake intrusions for males, because 20–50% of young are the result of EPCs. There are winners and losers because some males father many extra-pair young in addition to those in their own nest, but other males father no young at all despite having a social mate and feeding the young in her nest (Stutchbury *et al.* 1994, 1997). Neighbors are your worst enemy!

Breeding territories are set up nearly simultaneously within a population in the temperate zone spring, whereas, most tropical territorial setting up is not synchronous. Less appreciated is the fact that many of the differences in territory defense between temperate zone and tropical passerines are due to influences of extra-pair behavior. In temperate zone passerines, intrusions are a tactic of extra-pair mating systems. Male territory defense is typically much more vigorous than female perhaps because females face a greater time tradeoff between reproduction and territorial behavior (Elekonich 2000). The territorial system of tropical passerines is based on real estate, because EPCs are so uncommon (Chapter 4).

EPCs can be accomplished in a few minutes, so male vigilance against intruders must be high and persistent. EPC competition means that intrusion rates are high, and often escalate into fights. In Hooded Warblers, 20% of covert intrusions resulted in chases or fights with the territory owner (Stutchbury 1998b). A typical territory owner faces about one covert intrusion per hour and is involved in at least several chases or fights per day with intruding males! Even without the aid of radiotelemetry, such EPC chases are conspicuous and commonly seen in a wide variety of forest birds.

In tropical passerines actual intrusions onto territories are much less common (Freed 1987, Greenberg and Gradwohl 1997). Intense fights, when observed, usually involve take over attempts by juveniles or floaters, rather than interactions between neighbors. Border disputes, where single birds or pairs sing at boundaries are common but these rarely escalate into fights. Pairs of Checker-throated Antwrens, *Myrmotherula fulviventris*, for example, defend borders of their mixed species flock territory against conspecifics with loud and continuous *cheh-cheh-cheh* etc. calls accompanied by lateral body movements in rhythm with the calls. These clashes may last for many minutes.

Male song output is very high in most temperate passerines. It functions in extra-pair mating competition as well as establishing territory boundaries and attracting a social mate (see Chapter 6). In Hooded Warblers, males spend about 50% of their time singing even after they already have a social mate (Wiley *et al.* 1994, Stutchbury 1998b). Females assess the quality of potential extra-pair partners by assessing their singing output and, in some, song variability (Kempenaers *et al.* 1992, Hasselquist *et al.* 1996). Mate choice continues long after a male attracts a social mate to his territory, so males must maintain a high song output. In most tropical passerines song output is relatively low, even during the short dawn chorus (see Chapter 6). Many species are heard only between 0615 and 0730 or for even shorter periods. Some species have special dawn songs given by males only at or before dawn (Staicer *et al.* 1996).

Extra-pair mating also has an impact on territory settlement

patterns. Females prefer to settle on territories where extra-pair males are nearby (Wagner 1993, 1998, Wagner et al. 1996). This can result in a clustering of territories with females avoiding settling on territories isolated from potential extra-pair mates. This is clearly seen in the Least Flycatcher, *Empidonax minimus*, which has conspicuous clustering of tiny territories ('colonies') that shift in location from year to year (Wagner 1998, S. Tarof, in prep.). The high cost of obtaining EPCs from isolated territories may explain why so many forest passerines occur at low frequency in forest fragments that are big enough to support a breeding pair, but nevertheless are unoccupied (Morton 1992, Norris and Stutchbury 2001). We do not expect to see this EPC-based territory settlement pattern in tropical passerines.

Testosterone and territoriality

Testosterone is thought to be the key proximate mechanism driving territory defense in birds. Numerous studies of temperate zone birds have shown that testosterone level is high during the breeding season, especially during territory establishment and courtship in spring (Wingfield et al. 1990). Circulating levels of testosterone correlate with the individual's current state of aggression only when social relations are unstable, such as when territories are being established, when the mate is sexually receptive and when males mate guard (Wingfield et al. 1999). Testosterone levels increase dramatically in individuals challenged by intruders. Individuals with testosterone implants increase their territory size (Ketterson et al. 1992). Outside the breeding season, when gonads are regressed, individuals do not increase testosterone levels in response to challenges (Wingfield 1994). Wingfield and Hahn (1994) show that territorial aggression can be 'activated' in the absence of testosterone in a sedentary population of Song Sparrows, *Melospiza melodia*. However, they insist that persistence of aggression in the face of a 'simulated territorial intrusion' is still testosterone-dependent (Wingfield et al. 1999).

Tropical birds break all these rules. Many tropical birds have very low testosterone levels all year despite being highly aggressive and territorial (Dittami and Gwinner 1990, Levin and Wingfield 1992, Wingfield et al. 1992, Wingfield and Lewis 1993). In Panama, Spotted Antbird testosterone levels (and gonad size) remain very low year-round even though individuals defend permanent territories (Figure 5.2; Wikelski et al. 1999a). The seasonal pattern of testosterone is in sharp contrast with a temperate zone bird like the Red-winged Blackbird, *Agelaius phoeniceus* (Figure 5.2) or White-crowned Sparrow,

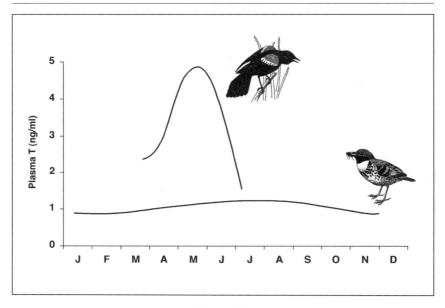

Figure 5.2
Plasma testosterone level (ng ml⁻¹) of male Spotted Antbirds in Panama over the year, including the breeding season from April–November (from Wikelski *et al.* 1999a), and for Red-winged Blackbirds in North America that breed from April–June (Johnsen 1998). Drawings from Medsger (1931) and Wetmore (1972).

Zonotrichia leucophrys (Hau *et al.* 2000). Testosterone in Spotted Antbirds became elevated only after prolonged challenges (> 90 min of playback experiments), whether or not gonads were enlarged, but even then the maximum ever recorded was 1.5 ng ml⁻¹. This is much lower than testosterone levels typical of temperate zone birds (4–6 ng ml⁻¹; Hau *et al.* 2000). The White-browed Sparrow Weaver, *Plocepasser mahali*, in Zambia has a similar low level of testosterone year-round, but in this species testosterone levels remain low even after simulated intrusions (Wingfield and Lewis 1992) and experimentally induced male takeovers (Wingfield *et al.* 1992).

Either tropical birds do not need testosterone for song and aggression, or they are highly sensitive to very low levels of testosterone (Hau *et al.* 2000). Testosterone implants did increase song and aggression in captive Spotted Antbirds, and physiological blocking of testosterone resulted in lower song output and aggression (Hau *et al.* 2000). This experiment shows that testosterone can affect song and aggression in tropical birds, though the T implants increased testosterone levels to about 6 ng ml⁻¹, much higher than is observed naturally. Such high

levels of testosterone, seen routinely in temperate zone birds, carry a high price in terms of immunosuppression and other trade offs (reviewed in Folstad and Karter 1992, Wingfield *et al.* 1999). Tropical birds can clearly defend and maintain territories without high testosterone levels, and in some species males can elevate testosterone opportunistically after prolonged territorial challenges (Wikelski *et al.* 1999a). Such 'social instability' occurs at low frequency in tropical territorial systems, largely because males are not competing with each other for extra-pair copulations (see Chapter 4). Without the strong sexual selection from extra-behavior there would be no benefit to maintaining elevated testosterone levels. Females also sing and defend territories, but nothing is known about the role of testosterone, or other hormones, in females (Hau *et al.* 2000).

Resource holding potential

Another tropical/temperate difference concerns the resource holding potential (RHP) of territory owners. Game theory models of territory defense distinguish RHP, the physical ability to defend and fight, from resource value (RV) which is the motivation to fight owing to the value of the territory to the contestant (Maynard Smith and Parker 1976). Together, RHP and RV help to predict asymmetries between contestants and therefore the outcome of contests as well as the likelihood of escalation. These models of territory defense are usually studied experimentally by examining the ability of removed owners to regain their territories from replacements. Temperate zone removals show that the probability that replacement individuals will defeat a former resident increases with replacement time (Krebs 1982, Beletsky 1996). An absence of 48 h is enough to tip the balance in favor of the replacement, who is usually a floater rather than a former territory owner. Only 16% of Red-winged Blackbird males regained their territory after being detained for 6–7 d, compared with 91% success in two-day removals (Beletsky and Orians 1987, 1989).

Few experiments like this have been performed in tropical birds. The value of breeding territories of temperate zone birds might differ greatly from the year-long territories of tropical birds. Dusky Antbirds of either gender always regained their territories, after they were released from captivity, regardless of replacement time up to 10 days (Morton *et al.* 2000). The replacements of the removed territory owners were other territory owners (not floaters), which switched mates and old territories for the new territory. The replacements, therefore, had the option of returning to the territory they emigrated from and ousting their replacements, if any. Removed residents, in contrast, did not have an alternative besides regaining their territory. The released owners' motivation to fight must have been higher than their replacements' motivation owing to this asymmetry.

5.3   Territory acquisition

In temperate species, removal experiments have shown that young birds of many species opt to delay breeding as non-territorial birds, either because no breeding positions are available (e.g. Stutchbury and Robertson 1987a, Stutchbury 1991) or in an attempt to gain a high quality breeding position much as cooperative breeders do (Zack and Stutchbury 1992). Tactics for gaining territories include wandering widely for vacancies (Stutchbury and Robertson 1987a), living secretly (Arcese 1989) or openly (Eckman 1988) on the territories of breeders, or evicting territory owners outright (Arcese 1987). For most tropical species removal experiments have rarely been performed (Morton 1977a, Levin 1996a, Morton et al. 2000) and we do not know whether floaters even exist, let alone how young birds obtain their first territory. Territory acquisition has been carefully studied in cooperatively breeding birds, many of them tropical, where young often inherit their natal territory or use their natal territory as a refuge from which to compete for breeding positions on nearby territories (Zack 1990). There are many similarities between cooperative breeders and other species in how young go about getting a breeding position (Zack and Stutchbury 1992).

Instead of wandering widely (e.g. 'floating' in the true sense) non-breeders in some resident temperate species gain a competitive advantage for breeding vacancies by associating closely with occupied breeding territories, via the same kind of site dominance advantage that territory owners enjoy (Birkhead and Clarkson 1985, Eckman 1988, Matthysen 1989). This tactic was first described in a tropical bird, the Rufous-collared Sparrow, *Zonotrichia capensis*, in Costa Rica (Smith 1978) where an 'underworld' of nonbreeders live furtively on the territories of breeders waiting for a vacancy to arise. Nonbreeders have well-defined home ranges that they defend from other nonbreeders, and when a breeder disappears the replacement is a nonbreeder whose home range included that territory. This kind of nonbreeder tactic for territory acquisition is likely to be common where year-round territory defense and high food availability make it possible for young birds to queue for a breeding position.

In tropical House Wrens, *Troglodytes aedon*, floaters of both sexes are transients and have at least two routes to territory acquisition (Freed 1986, 1987). Some wait passively for vacancies to occur, but adults are long-lived so vacancies arise rarely and are filled very rapidly. Floaters also attempt to evict territory owners instead of waiting for them to die, and kill their nestlings if the takeover occurs during breeding. Floaters sometimes form pair bonds, and then take over territories as a team.

In many monogamous tropical birds, juveniles live on their parent's territory for several to many months (e.g. Robinson *et al.* 2000, Morton *et al.* 2000), and this could give them an edge in competing for nearby vacancies that can arise any time of year. In Checker-throated Antwrens, juveniles live with their parents for a short time (1–2 months), but use this as a home base from which to challenge neighboring territory holders for ownership (Greenberg and Gradwohl 1997). Such challenges involve long contests of displays and chases, but rarely result in takeovers (unlike House Wrens). Territory acquisition is constrained by the specialized aerial dead-leaf foraging behavior of this species. Food resources are scarce within a territory owing to the limited availability of dead leaves, so young cannot live on their parents' territory for very long. However, living alone is risky because dead-leaf foraging makes it difficult to both search for food inside dead leaves and be vigilant for predators. This species joins mixed-species flocks to reduce the risk of predation while foraging. Most juveniles that were banded settled very close to their natal territory.

Spotted Antbird, *Hylophylax naevioides*, young also leave their parents' territory after only 6–8 weeks, perhaps owing to competition with adults at ant swarms. In this species, young birds do not settle near their parents' territory, and individuals that filled vacancies came from outside the study area on Pipeline Road, in Panama (J. Nesbitt, pers. comm.). Many non-territorial 'floaters' that were banded within the study site eventually acquired breeding territories there, suggesting a system similar to the 'underworld' described for Rufous-collared Sparrows (Smith 1978).

In Dusky Antbirds floaters are uncommon, probably because young birds live with their parents up until the next breeding season and reproductive success is very low (Morton and Stutchbury 2000, Morton *et al.* 2000). Many territories remained unoccupied after an occupant was experimentally removed during the nonbreeding season, or when their occupants disappeared naturally. This suggests there is no shortage of territories. New territories were established by pairs of juveniles, never by single birds, while 'widowed' adults remained on their territories and advertised for a new mate to join them (Morton *et al.* 2000). In Dusky Antbirds, territory establishment is likely constrained by predation. Pairs forage together in dense habitat and do not join mixed-species flocks, so living alone on a territory may expose a bird to a very high risk of predation by ambushing predators like vine snakes and boas. The quality of a territory, to a possible newcomer, may be greater if an experienced resident is present on it. Such a resident may be familiar with predators and their locations on the territory. We predict that, if one removes both residents from a territory, the quality of that territory will be reduced owing to the high cost of living alone, and it will remain unoccupied. These total removals can be compared with published data on replacement rates where only single individuals were removed (Morton *et al.* 2000) to test the 'experienced resident increases territory quality' hypothesis. We present this hypothesis to stimulate thinking about these tropical territorial systems. Year-long territoriality is, after all, the most common form of territoriality worldwide and we know almost nothing about sources of selection acting upon it.

How juvenile Dusky Antbirds join together to set up territories and how long they are tolerated on their parents' territories are not well known. They appeared to form pairbonds with juveniles on adjacent territories and use space contiguous to both parental territories. In other words, their territory was budded off a territory from tolerant parents. When we attempted to capture one such pair for banding, the mother of one of the paired juveniles left her territory and was captured! Perhaps she was 'helping out' the daughter.

Other year-round territorial passerines with juvenile retention differ in some details. Buff-breasted Wren, *Thryothorus leucotis*, removals resulted in 100% replacement either by banded young or neighboring adults, and sometimes unbanded floaters, usually within 24 h (S. Gill and B. J. M. Stutchbury, unpubl.). Floaters are uncommon, but do occur. All adults and their young were banded in this population, and occasionally unbanded birds were observed moving through territories or singing alone from a small area. In White-bellied Antbirds, *Myrmeciza longipes*, though, experimental removals often did not result in replacements, suggesting few floaters exist in either sex (B. Fedy and B. J. M. Stutchbury, unpubl.). These experiments have revealed a great variety among species that often occupy the same habitats, in terms of the frequency of floaters, how juveniles go about getting territories, and how far juveniles go from home to get breeding positions. Why are White-bellied Antbirds so different from Dusky Antbirds . . . we don't

know! This is so often the answer to questions we, and our students, pose about tropical birds.

5.4 Territory switching

Year-round territories have remarkably stable boundaries from year to year, even when owners are replaced (Greenberg and Gradwohl 1986, 1997). But permanent territoriality does not mean stasis. Several studies have found that territory switching occurs at a low rate in year-round residents (Willis 1974, Freed 1986, Greenberg and Gradwohl 1986, 1997, Woodworth *et al.* 1999, Morton *et al.* 2000).

But switching is an important aspect of territoriality, because the high adult longevity means that 25–50% of adults switch territories once during their lifetime (Greenberg and Gradwohl 1997, Morton *et al.* 2000). Pair bonds in Dusky Antbirds, apparently stable, are quickly broken when vacancies arise on nearby territories (Morton *et al.* 2000). We removed males or females from territories, then monitored who filled those vacancies and how quickly this occurred.

Figure 5.3 illustrates how vacancies on some territories are hotly competed for, while others are ignored. The male from territory A was replaced by an adjacent male in less than 12 h, which was in turn ousted by an unbanded male. We then removed the unbanded male and the same adjacent male moved back to reclaim the vacancy; he was replaced on his former territory by a yearling neighbor male within 12 h. When the original owner of A was released (after 168 h in captivity) he reclaimed his territory. The adjacent male returned to his former territory, as did the yearling male. In stark contrast, a removal on territory D resulted in no replacements after 70 h, and a natural disappearance of the male on territory E left that female unmated for 2.5 months!

Territory switching was equally common and rapid among males and females (Figure 5.4). When the original owners were released they always won back their territory, and replacements went back to their former territory, in a domino fashion. On average an individual switched territories about once in its lifetime. However, some individuals remained for many years (up to 10) on the same territory, suggesting these territories were preferred. Territory switching by breeding adults is probably a common tactic for gaining higher quality territories in many tropical birds. Levin (1996a) also documented several cases of territory switching by males and females after doing removal experiments in the Bay Wren, *Thryothorus nigricapillus*. In a congener, the Buff-breasted

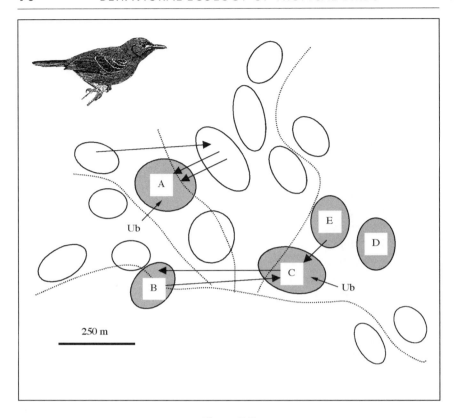

Figure 5.3
Outcome of male removal experiments (A–D) and one natural disappearance (E) in the Dusky Antbird during the nonbreeding season (Morton *et al.* 2000). Territories (solid lines) are located around the beginning of Pipeline Road, Soberania National Park, Panama. Dotted lines indicate roads. Arrows indicate source of replacements (Ub indicates replacement by unbanded male of unknown territory status). Drawing from Haverschmidt (1968).

Wren, about 15% of adults switch territories during their lifetime, though some pairs remain together on the same territory for over five years (S. Gill and B. J. M. Stutchbury, unpubl.).

How do adults decide to switch or not to switch territories? How do they assess territory quality on neighboring territories? Direct exploration of the territory is unlikely, and may be too risky if predation risk is high in unfamiliar terrain. Dusky Antbirds cannot easily leave their current territory to explore others because the mate they leave, even for a minute or two, will begin advertising for a new mate if its songs

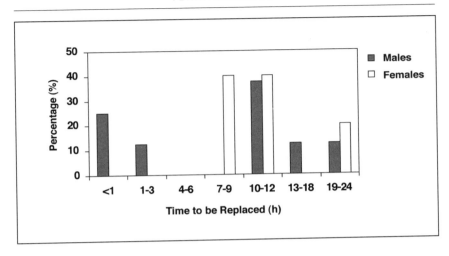

Figure 5.4
Frequency distribution of time to be replaced for male (n = 9) and female (n = 5)
Dusky Antbirds experimentally removed from territories.

or call notes are not answered (Morton and Derrickson 1996). Song output, especially during the dawn chorus, does not appear to be an accurate, if indirect, measure of food abundance on other territories but the area for foraging was greater in those territories that birds switched to than it was on those territories they left (Morton *et al.* 2000).

Mate abandonment is an important aspect of territory switching. Freed (1986) argued that permanent pair bonds had no intrinsic benefit in House Wrens, but rather were forced on individuals by the limited opportunities to switch territories. Switching territories usually means switching mates also, and the relative benefits to be gained from each remain unknown. From the practical perspective, it will be hard to tease apart mate choice from territory choice. In Buff-breasted Wrens some pairs have remained together, on the same territory, for over four years (S. Gill and B. J. M Stutchbury, in prep.). Is this because they are, respectively, high quality mates or because they both occupy a high quality territory? Perhaps the solution is to manipulate territory quality through food supplementation, to determine whether one can induce territory switching. This assumes that food availability is a key feature of territory quality, and we do not even know that that is true for tropical birds. Removal experiments have revealed a wide variety of outcomes, from rapid mate/territory switching in Dusky Antbirds and Buff-breasted Wrens to very little mate/territory switching in White-bellied Antbirds. A comparative approach to explain these differences among species is another way to determine how important mate and territory quality is for tropical birds.

6 | Communication

6.1 The assessment/management concept in communication

Communication and its study in animals means different things to different people. For most, communication revolves around the transfer of information (Owings and Morton 1998). A male who sings is sending information to his neighbors. Long tail streamers send information to females about male quality (Møller 1988, Andersson 1994). Readers will be familiar with such informational descriptions because this perspective has dominated the field. There is a 'sender' and a 'receiver'. Radios transmit information. Computers transmit information. Humans, because they have a language, can also transmit information when they speak. But animals do not talk (Morton and Page 1992). The information concept is anthropomorphic and does not do justice to the complex behavioral process that is communication. In this chapter we will not talk about which signals contain the most information or how signalers convey information to receivers. Instead we use a new perspective, the assessment/management concept (Owings and Morton 1998) to understand and discuss communication.

Animals communicate because they live in a social world and benefit from influencing the behavior of rivals, prospective mates, family members, predators, etc. Let's begin to understand the new perspective by looking at the sender and the receiver in a communication event. In the informational view senders are the key participant, it is thought, because they initiate communication by producing a signal containing information. In contrast, receivers perceive the signal somewhat passively then act accordingly, and in their own interest, depending on the 'information' received. But in the new perspective receivers, not senders, are the more important participant both in the proximate sense of immediate interactions and in the ultimate sense of how these cumulatively shape signals through natural selection. Receivers play the crucial role because they control the outcome of the interaction by how they respond, if at all, to the signal. Receivers determine what, if anything, has been accomplished by the signaler. In other words, the signaler is at the mercy of the receiver. Receivers do not merely absorb information that is sent their way, they determine, proximately and ultimately, what signals are used and how they are used. We refer to receivers as 'assessors' to highlight their important role.

Communication signals evolve not to convey information *per se*, but as a behavioral mechanism whereby a signaler can attempt to manage, or regulate, the actions of the assessor so that the signaler benefits from the interaction. An arms race of sorts, because what is best for the signaler is not necessarily best for the assessor (and vice versa). Bird song is an attempt to manage the behavior of another, for instance to keep an intruder from annexing a portion of a territory. A bird intent on trespassing into another's territory might ignore the defender's song if it perceived the singer to be far away. Assessors use sound degradation to estimate distance (see discussion on ranging below). An assessment by the would-be intruder that the defender is far away, and hence it is safe to intrude, becomes a source of selection on songbirds to use songs that transmit well and appear to be from a closer, more threatening, defender (Morton 1986, 1996a). A singing bird is not merely sending information about its location, it is attempting to manage the behavior of rivals by sounding as close as possible. It can only do this by modifying the way the degradation in its sounds are perceived over distance, because degradation is the feature by which distance is judged by assessors.

Consider a Dusky Antbird, *Cercomacra tyrannina*, pair foraging and duetting at the boundary of a neighbor's territory in their typical habitat of dense shrubs and grass at the edge of a tropical forest. The neighboring pair arrives in short order, gives a duet, and soon one perches close to the threatening pair, fluffs its back feathers (to look big) and gives a harsh low frequency 'growl.' The foraging pair quickly retreats. The signal of interest in this case is the growl given at close proximity to the rival. This is not simply a bird version of 'go away'. The signal given is low frequency and harsh *because* assessors will be intimidated by such signals. Animals assess fighting ability and physical threat based on size, and a virtually universal rule in communication is that low frequency sounds come from big animals (Morton 1977b). Thus, animals from a wide array of taxa use the same low frequency harsh 'growls' to intimidate, even humans. A person who says 'go away' in a low frequency harsh tone will sound much more intimidating than if one uses the exact same words 'go away' but in a high frequency tone

(Ohala 1984). Try it. The important point here is that the signals used in communication depend on how assessments are made; there is more to it than simply conveying information.

6.2   Song, territoriality and extra-pair behavior

Biologists interested in bird song need to realize that their temperate zone studies of female mate choice or song function are limited to 'breeding territory' systems (e.g. Møller 1991, Ratcliffe and Otter 1996, Searcy and Yasakawa 1996 and references therein). Defense of territories only during the breeding season typifies the vast majority of temperate zone passerines (Table 5.1). In decided contrast, breeding season territoriality is uncommon in tropical passerines and, therefore, for most bird species. Breeding territoriality was the focus of the earliest work on avian territoriality, including the work of Eliot Howard, Bernard Altum, Moffat, and Nice (Stokes 1974), because these biologists were confined to temperate latitudes. We need much more work on the other territorial systems found among tropical birds (Table 5.1) to begin to understand the evolution of territoriality and how communication is adapted to it.

Singing behavior in temperate zone birds is unique because, due to climatic constraints, song is highly correlated with territory establishment, pair formation, and breeding. The time devoted to pair formation and reproduction is a short pulse followed by a long non-breeding period during which birds stop or reduce territorial behavior, or migrate out of breeding ranges altogether. In sharp contrast, tropical territorial behavior and breeding seasons are long-term efforts (Baker 1938). Territories are often defended year-round, and no more intensively during the breeding than the nonbreeding season. Song is used in territorial defense throughout the year, and pair formation occurs infrequently and at any time of the year. Furthermore, singing and territorial defense are not entirely, or even largely, male behaviors. Other types of territorial systems are also common in the tropics (Table 5.1). Song is used year-long among pairmembers of 'fruit influenced' territorial species and mixed species flock members. Together with the standard year-long territorial species, these constitute 76% of the passerines in some tropical areas (Table 5.1). Clearly, the temperate zone model does not apply well to these species.

Recent examples continue to overgeneralize about the evolution of song based on data from breeding territorial systems. These reports, interesting in their own right, overgeneralize by insisting that bird song evolves largely as an intersexual form of communication based upon female choice. Møller *et al.* (2000) explain song repertoires of male birds as resulting from sexual selection because, they suggest, females prefer males with large repertoires because they suffer less from malarial parasites, so repertoires reflect health status. While these correlational studies are intriguing, their generality, if any, is limited to breeding territoriality. They are not generalizable to 'all' bird song because they do not encompass the diversity of song function in more common territorial systems. What about the 79% of passerines in Panama alone that have year-long pairbonds? Like the Carolina Wren, *Thryothorus ludovicianus*, in North America, it is not likely that song has much to do with female mate choice. Instead, this song functions for defending the territory both inside and out of periods of breeding (Morton 1996b). For them, the territory is essential for individual survival throughout the year, not only for reproduction during a portion of it.

Breeding territoriality contributes to latitudinal differences in the prevalence of extra-pair mating systems. It is due to this mating system that song function in temperate zone birds must be viewed as specialized and atypical (Chapter 4). In temperate latitudes, males devote time and energy competing to attract females for EPCs. Testosterone is the hormone of choice to facilitate the EPC competition amongst males (Chapter 5). Females actively seek EPCs from neighbors (Neudorf *et al.* 1997). If you are a male temperate zone bird your male neighbor is your worst enemy for you are more likely to lose paternity to neighbors than to more distant territory holders or nonterritorial floaters (Stutchbury *et al.* 1997, Stutchbury 1998b). Most tropical birds have not evolved extra-pair mating systems (Chapter 4) so their singing behavior is not influenced by competition for, and defense against, EPCs.

Song output varies dramatically between tropical and temperate zone birds because output is a key to females evaluating males for EPFs. High output, therefore, where a male might be forced to sing to the point of over-exertion (Morton 1986) is sexually selected. Figure 6.1 illustrates typical song output under an extra-pair mating system in typical temperate zone birds in contrast to two tropical birds, one with and one without extra-pair behavior. For most temperate birds with breeding territories (Mace 1987, Møller 1991, Pärt 1991, Krokene *et al.* 1996, Gil 1999) song peaks sometime between the time of pair formation and incubation, then falls as the breeding season progresses (Figure 6.1A). The same pattern of singing output occurs in a tropical

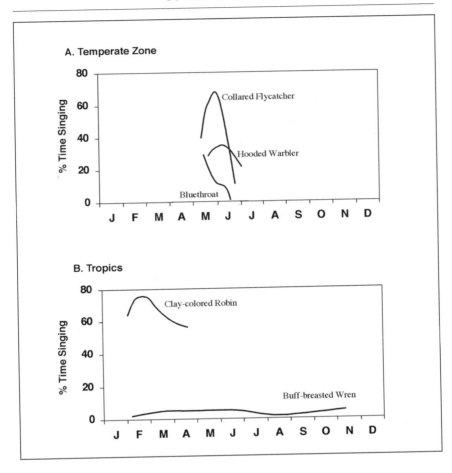

Figure 6.1

Song output (% time singing) versus time of year for A) typical temperate zone passerines the Collared Flycatcher *Ficedula albicollis* (Pärt 1991), Bluethroat *Luscinia svecica* (Krokene *et al.* 1996) and Hooded Warbler (Wiley *et al.* 1994) and B) the tropical Clay-colored Robin (Stutchbury *et al.* 1998) and Buff-breasted Wren (S. Gill, unpubl).

bird with breeding territoriality and EPFs, the Clay-colored Robin, *Turdus grayi*, but in a more typical passerine with year-long territories, the Buff-breasted Wren, *Thryothorus leucotis*, song output is very low and more or less invariable year-round (Figure 6.1B).

Song rate is very low for most tropical passerines, usually less than one song/minute, even for species that actively defend territories year-round (Figure 6.2; Wiley and Wiley 1977). The dawn chorus, which is

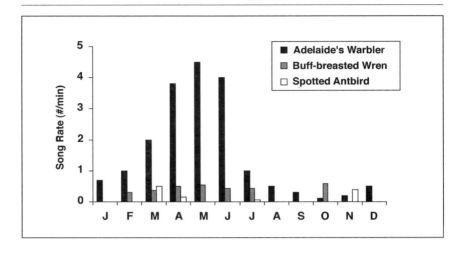

Figure 6.2
Variability among neotropical passerines in song rate and seasonality of the dawn
chorus. Data from Staicer et al. (1996), Wikelski et al. (2000) and S. Gill (unpubl.).

when song rate is often highest, still amounts to a paltry 0.5 songs min^{-1}
for Spotted Antbirds, *Hylophylax naevioides*, even during the breeding
season (Wikelski *et al.* 2000). White-bellied Antbirds, *Myrmeciza
longipes*, sing only several times per hour and do not increase song
output even at dawn (Fedy and Stutchbury, unpubl.). The dawn
chorus is impressive in some tropical birds, like the Yellow-bellied
Elaenia, *Elaenia flavogaster*, where males sing non-stop for some 15–20
min just before dawn (10–15 songs min^{-1}) during the breeding season,
but pairs sing only 10 times per hour during the daytime (Morton *et al.*
unpubl.). Adelaide's Warbler, *Dendroica adelaidae*, also have a pre-dawn
chorus given during the breeding season that peaks at 4–5 songs min^{-1}
(Figure 6.2), comparable to the song rate of many temperate zone birds
(Møller 1991, Pärt 1991, Titus *et al.* 1997, Gil *et al.* 1999). During the
day their song rate is much lower (0.1–1 song min^{-1}) and increases
slightly during the breeding season. Although song rates are typically
very low, for most tropical species singing increases dramatically in
response to playbacks or territorial challenges (Wiley and Wiley 1977,
Levin 1996b, Morton and Derrickson 1996).

If high song output can be used by females to assess males, then
output should reflect differences in male health or vigor and ultimately
in intrinsic male quality (good genes). This has been demonstrated in
several temperate zone species. For instance, female Blackcaps, *Sylvia
atricapilla*, use song rates rather than territorial quality per se in mating

decisions (Hoi-Leitner *et al.* 1995) as predicted earlier (Morton 1986). Variation in song output that could be used by females in such evaluations is illustrated by the Clay-Colored Robin (Figure 6.3). Differences in song output during the dawn chorus could be used by female robins in choosing males for EPFs (Stutchbury *et al.* 1998), just as for temperate zone birds.

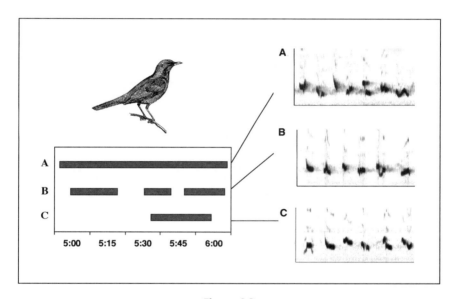

Figure 6.3

Singing patterns of three individual male Clay-colored Robins (A–C) during the predawn chorus (0500 to 0600) showing the timing and length of their singing bouts and sonograms showing their individually recognizable songs. Drawing from deSchauensee (1964).

The over-riding influence of extra-pair mating systems is seen in temperate species that are year-round residents (but do not defend territories year-round). In Black-capped Chickadees, *Poecile atricapillus*, social mate choice occurs gradually and months before breeding (Smith 1991). Still, females breed synchronously and often choose to copulate with nonmates (Smith 1988; Otter *et al.* 1994). Females assess males for copulation based on song output during the dawn chorus during the breeding season (Otter *et al.* 1997) and prefer males that begin singing earlier, sing longer, and have higher average and maximal rates. These are usually also males that are high ranking in winter feeding flocks (Otter *et al.* 1998) but females, apparently, do not remember those details and need a more proximate cue to a male's vigor.

Breeding territories are established nearly simultaneously by most temperate zone male birds. The competition amongst males is high. Neighborhoods of birds are highly unstable because territorial boundaries are unknown to the many newcomers to the neighborhood. Mate attraction is also going on, again more or less simultaneously, shortly after territories are established. Males are forced by female choice to sing a great deal, which is why food supplementation studies show a positive influence on song output, and a direct correlation between song output and female mate preferences, in the temperate zone (Table 6.1). Because they differ in all these respects, tropical birds, those without extra-pair mating systems, are predicted to sing little relative to birds of higher latitudes. Food supplementation studies are needed to test this.

Table 6.1
Studies showing the increase in song output following supplemental feeding.

Species	Reference
Pied Flycatcher	Alatalo *et al.* 1990
Ficedula hypoleuca	Gottlander 1987
Blackcap	Hoi-Leitner *et al.* 1995
Sylvia atricapilla	
Willow Warbler	Radesater *et al.* 1987
Phylloscopus trochilus	
Savannah Sparrow	Reid 1987
Passerculus sandwichensis	
Blackbird	Cuthill and Macdonald 1990
Turdus merula	
Carolina Wren	Morton 1982
. *Thryothorus ludovicianus*	Strain and Mumme 1988

Temperate birds might also sing more than tropical birds simply because they have more food. Carolina Wrens have year-round territories and do not have EPFs, at least in Alabama (Haggerty *et al.* in press). Not only do they respond positively to food provisioning (Table 6.1) but temperate populations sing much more than tropical ones (Figure 6.4). The temperate zone populations of this wren try to increase the size of their territories whenever vacancies arise. Those with larger territories have a better chance of surviving winter snows

(Morton and Shalter 1977). Perhaps males sing as much as possible so as to disrupt the foraging activities of neighbors to better the chances that vacancies will arise (Morton 1982). Food supplementation studies have not been performed on tropical birds to see if their song rates are influenced by food availability. We predict that their singing will not be influenced by food availability because they sing only to defend real estate and are not competing for extra-pair matings. Song output during the dawn chorus in Dusky Antbirds did not increase in birds whose territories were supplemented with mealworm feeders (Morton unpubl.).

6.3 Sex role convergence in song

One manifestation of these latitudinal differences is that singing in female birds is noteworthy in temperate zone birds (Gilbert and Carroll 1999) but common and characteristic among tropical species (Morton 1996b). It is clear why females of many tropical species sing; they are territorial year-round and vigorously defend their territory from other females. Tropical females are defending food resources for

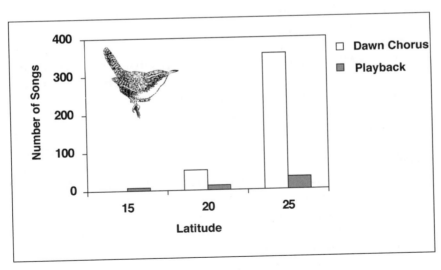

Figure 6.4
Song rate of male Carolina Wrens outside of the breeding season at different latitudes. Shown are the number of songs given per hour during the dawn chorus, and the total number of songs given after stimulus by a playback of conspecific song within the male's territory (data from Morton 1982). Drawing from Owings and Morton (1998).

themselves and, less importantly for their young, on a permanent basis or at least for a relatively long time. This favors singing in females as well as males, not to defend shared territories as a team, but to defend against competitors of the same sex. Duetting, wherein pair members sing together, has less to do with the pairbond itself than in insuring that aggression is used against competitors of the same gender (Farabaugh 1982). Tropical females can sometimes maintain a year-long territory even without a mate (Morton *et al.* 2000). For them, as in males, songs are better than call notes to manage assessor behavior over large areas. Sex role convergence occurs due to a need for females, as well as males, to defend territories (Levin 1996a,b, Morton 1996b).

Female song occurs in some temperate zone species, but is not always territorial in function. Female Eastern Bluebirds, *Sialia sialis*, sing to induce their mates to mob nest predators (Morton *et al.* 1978) and female song in other species is associated with coordination of parental care (Ritchison 1983, Halkin 1997). Females, however, generally don't sing in the temperate zone. The reasons for this lack of song are unknown. Territorial defense certainly occurs in temperate zone species, and females respond to playbacks of female call notes and attack female mounts (and live intruders!). So why are call notes used for territory defense instead of song? Singing is costly, and the stakes are likely lower for temperate females, which defend their territories for a period of only several weeks. Female defense likely revolves around preventing settlement by additional females, so that male parental care for feeding young is guaranteed. Contrast this with tropical females, who defend critical food supplies that enable them to survive year-round. Because nesting is synchronous in temperate areas, females can accomplish territory defense for their short breeding season using cheaper call notes without the need for the species- and individual-specificity of song that contributes to male success in extra-pair competition. Another advantage of using calls, rather than song, is that nonsinging females do not devote costly brain space to song production and perception (Brenowitz *et al.* 1995, Nealen and Perkel 2000).

The functional role of calls in territory defense is clearly seen in migrants that defend nonbreeding territories in the tropics. Although the nonbreeding territories are separated in space from their breeding territories, these species should rightfully be considered to have year-long territories. A major difference from their breeding territories, however, is that both genders defend nonbreeding territories against all conspecific individuals. Kentucky Warblers, *Oporornis formosus*, for example, have typical breeding territories, with males singing and setting up territories, attracting females, and everyone pursuing EPCs. Later they also defend individual territories during migration and at overwintering sites in tropical forests (Mabey and Morton 1992) where species-specific call notes are used to regulate the behavior of all conspecifics. Neither gender uses song in defense of nonbreeding or migration territories. The Kentucky Warbler system is used by about 31 species of nearctic migratory passerines, which have both breeding territoriality while in eastern North America and then permanent territories, while they are in their tropical homes, as do so many other tropical birds (Morton 1980, Rappole 1995).

In a few temperate species, females do sing to defend nonbreeding territories but not on their breeding territories, where song is restricted to males. Song may be favored owing to its long distance propagation qualities over call notes in these species. The White-eyed Vireo, *Vireo griseus*, and European Robin, *Erithacus rubecula*, are examples. The vireo requires a source of fruit from the gumbo limbo tree, *Bersera simaruba*, on its territory (Greenberg *et al.* 1995). The tree, of course, 'wants' its fruit eaten so the vireo may be forced to threaten distant conspecifics to keep them away from the visually obvious attraction of the fruit. Song would do this better than call notes, which may not carry far enough.

6.4 Song ranging, neighborhood stability and dialects

Ranging was discovered accidentally. We were trying to capture White-breasted Wood-wrens, *Henicorhina leucosticta*, in 1976 along the Pipeline Road in what is now Parque Nacional Soberania in Panama. Our goal was to release them on Barro Colorado Island, where these wrens and other species disappeared from the Island for unknown reasons. We were looking for the causes of their extirpation (Morton 1978). One way to capture pairs of birds having year-long territoriality is to play back a tape recording of their songs, making sure to include duets or examples of both gender's songs, near a mist net placed within their territory.

We used a high quality recording of duetting White-breasted Wood-wrens for playback. It came from Cerro Campana, only about 30 km from the Pipeline Road. Oddly, only male wrens were attracted to the playback. The males did not react particularly strongly but we were able to capture enough for the release. When we recorded a local version of the male song, Pipeline Road males responded much quicker and more vigorously. Females never showed themselves nor did they utter a

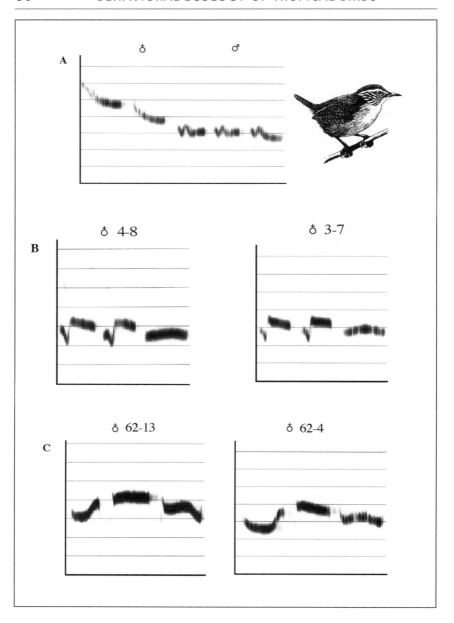

Figure 6.5
Sonogram of the A) duet of a White-breasted Woodwren pair B) a song type shared by different females in the same Pipeline Road population and C) a different song type also shared by females in the Pipeline Road population. Drawing from Wetmore (1984).

sound in response. They completely ignored the duet playback from Cerro Campana though it contained as much female as male song. We thought that females must not sing in this population and tucked the observation away for future study.

A few years later we discovered that female wood-wrens do indeed sing in the Pipeline Road population (Figure 6.5)! By chance, the male song version on our Cerro Campana playback tape was similar to one shared by the local male wrens but the female song was unlike any sung by the Pipeline Road females. Furthermore, we found that male wood-wrens each have a large repertoire of ca. 30 songs, most of which are not identical to songs of neighboring males. In great contrast, female wood-wrens have only 4 or 5 songtypes and all of these are shared with neighboring females (Figure 6.5). Male wood-wrens, then, have repertoires of songs, mostly unique to individual males (unshared) and females have a small repertoire of songs that form a dialect. Repertoires in males, dialects in females have not been described before in songbirds.

We use the term *dialect* to refer to contiguous populations with clearly differentiated vocal patterns, including song neighborhoods (such as described for Indigo Buntings, *Passerina cyanea* (Payne 1983)) but not microgeographical variation (Krebs and Kroodsma 1980), such as that described for the Carolina Wren (Morton 1987). This wren is typical of most species with repertoires in that neighboring males share a large percentage (ca. 85%) of their songtypes, but not all of them, and the percentage of songs shared by males decreases linearly with distance (Morton 1987). Repertoires and dialects are not mutually exclusive but there are no known cases of species with more than about eight songs having all birds in a contiguous population share all of them as a dialect 'repertoire.' Instead, one sees a gradual decrease in the percentage of the songs in the repertoires shared among neighbors with increasing distance between the birds being compared.

The response by male wood-wrens to the Cerro Campana song and the total lack of response by females to it puzzled us. We began to see a pattern in other species between responses to playbacks in populations or species having dialects versus those having large repertoires. Birds with dialects respond strongly only to their local dialect. Birds with repertoires respond to any conspecific song. Why? It wasn't until we read Doug Richards work on Carolina Wrens that we knew the likely answer. The answer led to a general theory of bird song evolution, the ranging hypothesis.

Richards (1981) discovered that the key to ranging was the ability of a listener to use song degradation to estimate its distance from a singer. Carolina Wrens virtually ignored degraded songs but responded strongly to these same songs when undegraded versions were broadcast at the same volume. How did they do this? We suggested that the assessor must have memorized the song so that it is able to match the song it hears with the undegraded version it has in its own memory (Morton 1982, 1986). In this way it can assess the amount of degradation in the song and, because degradation increases with distance, it can estimate its distance from the singer. It matters little if the song is from a neighbor or a stranger: as long as the physical structure of the song is memorized, the listener can range its distance from the singer no matter who sings it (Falls et al. 1982).

Ranging ties together a lot of 'loose ends' about bird song and its function and evolution. For example, it explains why birds have evolved an unusual degree of temporal resolving power even though their ability to discriminate frequency and intensity changes is not noteworthy. A budgerigar can resolve events happening faster than once every 1.2 msec whereas humans lose sensitivity to events happening faster than once every 5 or 6 msec (Dooling 1982). Birds are comparable to echo locating bats with respect to time resolution (Konishi 1969). Temporal resolving power improves a bird's ranging ability because it allows them to hear echoes from tree trunks and branches that contribute to degradation in a song as it travels farther and farther from the singer. Amplitude is probably not useful for ranging because it varies too much. A singing bird that turns its head away from a listener could appear to be farther away when, in fact, it hasn't moved, if amplitude was very important to distance estimation. On the other hand changes in the frequency mix of a song might also be useful for ranging (Naguib 1995, Fotheringham et al. 1997).

No wonder White-breasted Wood-wren males did not respond as strongly to the Cerro Campana song! They were unable to range the Campana song playback because they had no song in their own memory to judge it against. But because the males have repertoires, largely unshared, they responded to the species-specific qualities of the song. Their response was much stronger once we played back local songs that added degradation assessment to species-specificity. These local songs were more threatening and evoked strong territorial aggression in the males. Female wood-wrens, with their dialects, did not respond at all and we will discuss why in a moment.

First, what are dialects and why are dialects favored in some species, but repertoires (and no distinct dialects) favored in others?

Temperate/tropical differences in the function of song in territory defense and mate attraction is closely associated with differences in repertoire size, individual distinctiveness, and dialects. Tropical species tend to have stable neighborhoods with long-standing territory boundaries known by neighboring territory-holders, and very little active mate attraction (either social or extra-pair). Temperate species tend to have unstable neighborhoods with new boundaries established each year with many new neighbors to compete with, and a very active and persistent period of mate attraction (first social, then extra-pair).

Ranging theory can explain the effect of neighborhood stability and mate attraction on repertoire size and dialects. First, we should point out a major difference in passerine birds that influences songs. Oscines are often said to 'learn' their songs, because their mature songs are acquired partly through the experience of hearing them whereas, because non-oscines do not require such experience, their songs are described as innate. Obviously such dichotomous terms are shorthand for the complex developmental interactions that occur during song acquisition (Marler 1999) but for our purposes this dichotomy is useful. Oscines, because they learn their songs, can acquire repertoires with variable amounts of sharing within neighborhoods of competitors, but many species sing instead but a single song. These single songs might be completely shared by birds within a population as a dialect, or be individually distinctive with no dialect. 'Shared' means that one or more birds sing one or more songs having the same physical structure such that they sound the same and their spectrograms are identical, or almost identical. Non-oscine passerines (antbirds, flycatchers, etc.) do not learn their songs and, consequently, all individuals share the same song or songs, regardless of whether they have a single song, or a small repertoire of songs. Non-oscine passerines, which cannot be said to have dialects because they do not learn songs are, nevertheless, similar to birds with dialects because neighbors share songs.

The reason some oscines develop single songs and dialects, rather than flouting their song learning ability by acquiring repertoires, lies in the stability of their neighborhood. By neighborhood we mean a small population of birds that can actually hear one another as they defend territories. The birds interact vocally and try to manage and assess each other through communication instead of the alternative to communication, namely fighting. By neighborhood stability, we mean that the membership in the group of birds that can hear one another changes relatively little over time.

Neighborhood stability favors the use of songs that can be easily ranged by others. Such songs would be those shared with other birds because these can be ranged most accurately. Such distances can be large or small, for even birds breeding in dense colonies use shared songs to threaten one another. Dominant male Yellow-rumped Caciques, *Cacicus cela*, converge in their songs within colonies (Feekes 1981) as do temperate male Purple Martins, *Progne subis*, sitting only centimeters apart. A shared song is, by definition, in your neighbor's memory and, when it hears it, it knows two things: where you are in relation to the territorial boundaries you both share and where you are in relation to its own geographic location. Where you are in relation to the territorial boundary is only germane where the neighborhood is stable and rivals have long-established boundaries that everyone knows. Here, indicating your distance to a rival works to regulate its behavior when it already knows the territorial boundary you share. Where stable neighborhoods are usual, the songs are often dialects.

Neighborhood stability is characteristic of the tropics due to long-standing, nearly unchanging territorial boundaries and low turnover in adults (Greenberg and Gradwohl 1986, 1997, Morton *et al.* 2000). Dialects are very common in tropical birds but have not been well studied there. In Panama, near the town of Gamboa, the Chagres River is wide enough that birds cannot hear conspecifics on the other side. Dialect differences can be detected in many species that live on both sides. Blue-black Grosbeaks, *Cyanocompsa cyanoides*, have distinctive dialects (Figure 6.6). These are fairly widespread geographically with the 'Gamboa' dialect encompassing the Pipeline Road area as well as Gamboa, a linear distance of about 15 km at a minimum. On the other side of the Chagres River, the 'Old Gamboa Road' dialect also occurs along the road through the Parque Nacional Soberania, formerly known as Madden Forest. On Barro Colorado Island, the same distinctive dialect has existed, unchanged, at least since 1964 (Morton, pers. obs.).

Many tropical bird dialects have large geographical ranges such that the species appears to have but one song for the entire population regardless of geographic boundaries. Amongst these are saltators, *Saltator* spp., vireos, *Hylophilus* sp., *Smaragdolanius* sp., *Vireo* sp., and various emberizine finches. It is possible that tropical oscines such as these learn songs in a different way compared to species that exhibit local dialects.

Non-oscines, common in the tropics, do not learn their songs and so cannot be said to have dialects. But because songs are innate, song sharing among neighboring conspecifics is the rule. For example, in the Dusky Antbird, the only non-oscine passerine whose ability to range

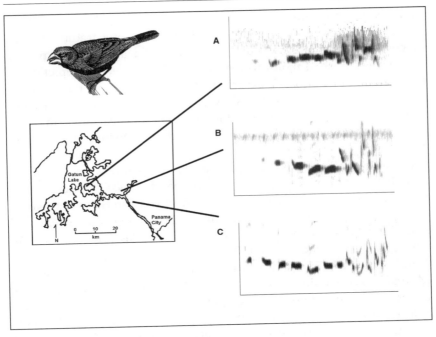

Figure 6.6

Dialects of the Blue-black Grosbeak in the canal area of Panama. Drawing from Skutch (1954).

has been studied, both genders are able to range (Morton and Derrickson 1996). As with all non-oscines, antbirds use innate shared songs, just as oscines use dialects or shared songs, to provide 'honest' distance cues to listeners. Dusky Antbirds have stable neighborhoods with unchanging territorial boundaries (Morton *et al.* 2000). The consequence of this stability is that their innate songs are *functionally equivalent* to dialects in protecting their territorial boundaries. Thus either innate songs or learned dialects are perfectly adequate for managing behavior in stable neighborhoods, which describes the situation for most tropical birds, whether they learn songs or not.

In some temperate migratory species, dialects are found but on a smaller geographic scale, at the level of a few interacting males. These are called song neighborhoods and arise due to the learning of the local song by yearling males attempting to settle for the first time (Payne 1983, 1996). The difference between a song neighborhood and a dialect is only one of scale: dialects are usually of a larger geographic scale than song neighborhoods. But even within a dialect variants occur within smaller neighborhood groups of interacting birds (Baptista 1975). Some call these neighborhood dialects 'song traditions' and suggest that they represent 'cultural evolution' because they are passed on from established males to newcomers. In reality, the depiction of this situation as 'culture' is describing their continuation, but not their function or the reason they exist. 'Culture,' as it is sometimes used in relation to bird songs, is an epiphenomonon of song learning and is distracting because it has no real scientific utility (Owings and Morton 1998).

Stable neighborhoods also occur in the temperate zone, usually in areas lacking very cold and snowy winter conditions. Climates such as these are termed 'Mediterranean' and occur in western coastal North America. Characteristic birds include Wrentits, *Chamaea fasciata*, and nonmigratory White-crowned Sparrows, *Zonotrichia leucophrys nutalli*, which have pronounced and well studied dialects (Marler and Tamura 1962, Baptista 1975). Widespread species, such as the Song Sparrow, *Melospiza melodia*, illustrate the relation between song sharing among males and neighborhood stability quite well. In stable populations near Seattle, Washington, males share many of their song types and usually match the songs of rivals in song duels. Those males that share most songs with neighbors have a longer lifetime tenure on their territories (Beecher *et al.* 2000a,b). In highly migratory populations of Song Sparrows, little or no song sharing or matched-countersinging occurs (Harris and Lemon 1974). Sedge wrens, *Cistothorus* sp., illustrate the relation of neighborhood stability and song sharing over a broad latitudinal range. Tropical species have dialects and stable neighborhoods, whereas the North American representative, whose breeding habitat is ephemeral from year to year, develops individually-specific songs because their neighborhoods are extremely unstable (Kroodsma *et al.* 1999). The reason these relationships hold is due to the differences we have described in the ways managers use ranging to control assessor behavior.

Carolina Wrens, even though they have year-long territories, are fairly typical of temperate zone birds with high turnover as a result of winter mortality. Males share about 85% of their repertoires of 30+ songs with at least one of their several immediate neighbors (Morton 1987). Neighbors use shared songs to countersing with neighbors over territory boundaries. The shared songs, because they can be accurately ranged by neighbors that sing them too, function to threaten neighbors over common territorial boundaries. Matching songs with neighbors while countersinging with them uses their ability to range to manage them without the energy-demanding need to approach them directly (see also Krebs *et al.* 1981). But, singing birds don't always 'want' to be ranged and this may explain why unshared songs are important. The unshared songs must be more disturbing to neighbors because they cannot range them accurately. These songs might be likened to the sound of an incoming mortar round: you're not sure where it will strike but you had better pay attention! The more vigorous a singer is, the more time he has devoted to singing, a tradeoff with foraging time (Strain and Mumme 1988). Carolina Wrens always try to increase the size of their territories because those with the largest territories are more likely to survive winter snows. Larger territories have more windfalls that keep snow off the ground leaf litter where the birds hunt for invertebrate food (Morton and Shalter 1977). Thus Carolina Wrens have very unstable boundaries because they change often. Singing to disturb neighbors might help to weaken them if they are forced to stop looking for food and investigate the whereabouts of a trespasser (Morton 1982).

Extra-pair mating systems also affect dialects because they favor individualistic songs. Extra-pair mating systems favor individuality in singing because females can track the singing output of specific males and choose them for EPFs. The North American sedge wren's individually distinctive song is shared with another widespread species, the White-crowned Sparrow, *Zonotrichia leucophrys*. White-crowned Sparrows are famous for having dialects (Marler and Tamura 1962) in populations with stable neighborhoods in western North America. The migratory eastern North American race, however, has individually distinctive single song repertoires (Austen and Handford 1991). The prediction is clear but as yet untested: extra-pair mating systems should characterize the populations with individualistic songs whereas resident populations with low levels of extra-pair mating should characterize populations with dialects. It follows that tropical birds should have dialect song systems both because their neighborhoods are very stable and because they do not have extra-pair mating systems that promote individuality in song.

In the extra-pair breeding systems characteristic of species with breeding territories, territory establishment and female choice may favor temporal changes in song function during a breeding season. Let's examine how ranging, and trying not to be ranged, functions as social conditions change during territory settlement in migrants. Many species of temperate zone paruline warblers switch from 'repeat mode' singing, wherein they use an individually distinctive song, to 'mixed mode' singing, which involves using larger numbers of songs from their repertoire that are shared amongst the males. Females should be better at ascribing song output to a particular male if he sings one individually distinctive songtype and it follows that repeat mode is known to function in attracting a social mate. Mixed mode singing has a very different function that is effective once boundaries are well established. Mixed, or dialect mode, threatens neighboring males to reduce their attempts to intrude for the purpose of obtaining extra-pair copulations. Perhaps female warblers have made their extra-pair mating decisions earlier and now the balance has swung towards using matched countersinging to keep other males at bay. In Hooded Warblers, for example, males leave their territories and intrude on neighbors nearly once per hour in pursuit of EPFs, often egged on by resident females (Chapter 5). Like the paruline warblers, Bush Warblers, *Cettia cetti*, in Korea have separate songs for male–male competition and for attracting females (Park and Park 2000). Catchpole and Leisler (1996) showed that complex songs are attractive to females and two simpler songs are used in male–male aggression in the Aquatic Warbler, *Acrocephalus paludicola*, but they did not mention whether or not the gender-specific songs differed in individuality.

The contrasts between temperate and tropical avian mating and territorial systems, especially more knowledge of the highly diverse tropical systems, should foster comparative studies of how biological differences in assessors are related to the diversity of singing by managers. We have already seen how female and male song, sex roles, and dialects are related to neighborhood stability, and extra-pair mating systems. Predictions can be generated and tested. The White-breasted Wood-wrens, where males have repertoires and females shared dialects, present a fascinating example. Do males turnover more rapidly than females producing both unstable *and* stable neighborhoods within the same species? Or, do females use male song diversity for choosing mates, which would favor repertoires in them?

Repertoires of songs are probably generally thought to arise due to sexual selection. This judgement is premature because repertoires have not been adequately studied in the plethora of tropical birds with year-long territories. The Carolina Wren is probably typical of most tropical birds. Its distribution runs from Guatemala, through the Yucatan Peninsula to Maine and, throughout this range, it retains year-long territoriality. It has a large repertoire, as mentioned above, but uses a close contact call that sounds like *tsuck* when courting a female. Furthermore, a wren can determine the gender of another individual only through nonsong vocalizations (Owings and Morton 1998, p. 235). It does not use song except for male–male interactions. Certainly song is important in mate choice for most birds with breeding territoriality in the temperate zone. But will these studies of repertoires in birds with breeding territories, which highlight female mate choice and song as an indicator of male quality, provide an adequate general explanation for song function and the evolution of learning in oscines (e.g. Nowicki *et al.* 1998)? The answer will come only after more study of repertoires, mate choice, and song function in the other territorial systems found in tropical birds.

6.5   Plumage signals, territoriality and extra-pair behavior

Conspicuous plumage signals generally result from sexual selection, either intrasexual competition or mate choice (reviewed in Andersson 1994). Plumage manipulation studies, a powerful tool for studying the adaptive significance of particular plumage features, show that plumage signals can be effective in managing receivers during territorial interactions (Peek 1972, Smith 1972, Rohwer and Roskaft 1989, Quarström 1997, but see Stutchbury 1992) and mate choice (Møller 1988, Hill 1990, Johnsen *et al.* 1998b). For instance, masking the red epaulet of Red-winged Blackbirds, *Agelaius phoeniceus*, results in males losing their territory altogether, suffering increased intruder pressure, and having smaller territories (Peek 1972, Smith 1972). Male Barn Swallows with lengthened tails obtain social mates earlier (Møller 1988) and obtain more extra-pair matings (Møller 1992, Saino *et al.* 1997b). These studies have been done with temperate zone species that have extra-pair matings and defend breeding territories for only a short time. How important are plumage signals in territory defense and mate attraction for tropical birds that defend year-round territories and have few opportunities for choosing mates?

Given that territory boundaries are usually stable and are not redrawn annually in the tropics, it is not so clear how important plumage signals are in territory defense. Most territorial tropical passerines studied to date have very low intrusion rates and a low frequency of border disputes relative to temperate zone birds (Chapter 5) and low song rates so it is quite possible that plumage manipulations on territory owners would have little or no immediate effect. Similarly for mate choice, because individuals so rarely obtain new mates. However, territory defense is swift and vigorous despite the low frequency of intrusions, and the plumage signal could be critical for these relatively rare challenges.

Male and female Dusky Antbirds have a white backspot which is coverable. Individuals reveal this white backspot by raising the back-feathers, and this is seen during intrasexual aggressive encounters and playback experiments. One could test the effectiveness of this signal in managing the behavior of intruders using plumage dyes to mask the signal, as has been done for the coverable badges of Red-winged Black-birds (Peek 1972, Smith 1972). Similarly, one could quantify aggression toward mounts prepared with or without the backspot showing (e.g. Hansen and Rohwer 1986). These kinds of simple experiments have not been done with tropical birds.

Subadult plumage in yearling males is common in many temperate zone songbirds, and often functions in competition for territories (Rohwer *et al.* 1980, Lyon and Montgomerie 1986). Yearling males return later in the spring than older males, and are at a competitive disadvantage in the spring rush for territories. Young males sport a dull female-like plumage in their first year that signals their subordinance to territorial adult males, thereby reducing aggression while they are searching for and claiming breeding territories in the spring (Studd and Robertson 1985, Lyon and Montgomerie 1986). It does not seem to function via female-mimicry (Stutchbury 1991, Enstrom 1993). Most subadult males do attract social mates once they obtain a territory, but can suffer a high incidence of cuckoldry from older males because females prefer adults as genetic mates (Morton *et al.* 1990, Wagner *et al.* 1996). Since most of these species also have a subadult plumage in the wintering areas this dull plumage may be an adaptation for competing for winter territories or signaling subordinance in winter flocks, in addition to or instead of having a breeding season function (Rohwer and Butcher 1988).

Subadult plumages have not been well studied in territorial tropical birds. In some lekking manakins young males have delayed plumage maturation for five years which clearly functions in subordinance signaling to older displaying males (Foster 1987b, McDonald 1989). Dusky Antbird males have a subadult plumage for their first year (Morton and Stutchbury 2000), but there is good reason to think it may function differently in communication from what is typical for temperate zone birds with breeding season territoriality. Dusky Antbird juveniles live on their parents' territory until the next breeding season rather than 'floating' independently. In this and other species yearlings use their home territory as a base from which to monitor, and fill, nearby vacancies (Chapter

5). The subadult plumage is presumably part of a general delayed dispersal strategy (Zack and Stutchbury 1992), but exactly how it benefits young males is unknown. Perhaps the subadult plumage is important for reducing predation risk while a juvenile (e.g. Selander 1965), or even signaling subordinance to fathers. We do know that widowed adult females sometimes 'reject' subadult male replacements by ignoring them and continuing to give courtship song despite the subadult male's attempts to duet with her (Morton *et al.* 2000).

The long lifespan of tropical birds should favor a delayed breeding strategy and subadult plumage even more so than occurs in temperate regions (Studd and Robertson 1985, Zack and Stutchbury 1992). Temperate zone migrants delay plumage maturation for only one year (Rohwer *et al.* 1980) but in manakins and some tanagers young birds do not attain adult coloration until they are 4 or 5 years old (McDonald 1989, Isler and Isler 1999). Female subadult plumage is uncommon in temperate passerines (Stutchbury and Robertson 1987b, Stutchbury *et al.* 1994), but should be more common in the tropics given the relatively equal sex roles in tropical passerines in terms of territory defense and reproduction. This is certainly true for tanagers, a largely tropical group, where subadult plumage occurs in about 50% of the species, most of which feature subadult plumage in both sexes (Isler and Isler 1999). Subadult plumages occur in a wide diversity of social and communicative situations. Both sexes have a distinct subadult plumage in the territorial and insectivorous Black-cheeked Ant-tanager, *Habia atrimaxillaris*, the largely frugivorous Bay-headed Tanager, *Tangara gyrola*, the White-shouldered Tanager, *Tachyphonus luctuosus*, which joins insectivorous mixed-species flocks, and the Black-faced Tanager, *Schistochlamys melanopis*, which is an omnivorous intratropical migrant. Subordinance signaling in a variety of social contexts (foraging, competition for territories, cooperative displays at leks) may be a general explanation for subadult plumages, but we still do not understand the selective basis for variation among species.

Extra-pair fertilizations create strong sexual selection (Stutchbury *et al.* 1997) and this could have a big impact on the evolution of plumage signals. Plumage manipulations in some temperate species have shown that females prefer extra-pair partners with plumage ornaments like long tails (Møller 1992, Saino *et al.* 1997b) and bright badges (Johnsen *et al.* 1998b). One could therefore predict that males should have more pronounced plumage brightness and sexual dichromatism in temperate than tropical species, due to the high frequency of EPFs. Sexual dichromatism is positively correlated with frequency of EPFs (Møller and Birkhead 1994), but the association is very weak and explains only 9% of the variance among species, even after controlling for phylogenetic effects. Certainly the absence of bright male coloration and dichromatism is no indication of sexual monogamy. The plumage of the Clay-colored Robin is as dull as one can get, despite a high EPF frequency (Stutchbury et al. 1998). Males have individually recognizable songs and some males sing virtually non-stop for an hour or more before dawn (Figure 6.3), so in this case sexual selection acts on vocal signals rather than visual ones. Dull monochromatic plumages are common even in lekking species despite very strong sexual selection (Trail 1990, Bleiweiss 1997). Despite strong sexual selection due to EPFs, we do not see brighter males and more dichromatic plumages in temperate zone species as a general rule (Bailey 1978).

Although sexual selection predominates as a force on plumage color in temperate areas, there are good examples where this is not the case for some tropical birds. Bright plumage in hummingbirds is associated with interspecific territoriality, not sexual selection, with dominant larger species being brighter than smaller subordinate species (Bleiweiss 1985). The dull and monochromatic plumage in lekking hermits is due to the similar subordinate foraging strategies of males and females that are non-territorial trapliners (Stiles and Wolf 1979). Similarly, in mixed species flocks 'nuclear' species are inconspicuously colored to facilitate flock formation and cohesion (Moynihan 1962).

Any discussion of plumage coloration in birds is complicated, to say the least, by the fact that we really do not know what birds look like! Birds see well in the UV range (Chen and Goldsmith 1986, Cuthill et al. 1999) but humans do not, and many species have plumage signals that reflect strongly in the UV range. For instance, male White-fronted Manakins, *Lepidothrix serena*, have a white patch on the front of the crown that has a reflectance (20–24%) in the UV range that is among the highest ever recorded for any animal (Endler and Théry 1996). UV 'badges', invisible to us, are used in mate choice (Bennett et al. 1996, Johnsen et al. 1998a, Hunt et al. 1999) and likely are also important for aggression within a sex. It would be convenient for us if plumage brightness and sexual dimorphism in the human-visible range correlated well with UV reflectance, but this is not so. Several species that appear sexually monochromatic are actually dichromatic in the UV range, and this is probably widespread in birds (Cuthill et al. 1999, Hunt et al. 1999). Why do birds use the UV range in signaling? It is not simply because they can see UV. Many mammals (including us!) cannot see well in the UV range, and presumably birds are taking advantage of being cryptic to predators but conspicuous to each other. It is unknown whether tropical birds differ in their use of UV signals in communication.

6.6 Sex role convergence in plumage signals

Latitudinal differences in sex roles of singing are clear; female song is vastly more common in the tropics. Such a broad generalization does not hold true for plumage signals, females are not necessarily more brightly colored in the tropics. Comparisons within families are most revealing from an evolutionary standpoint. Temperate zone wrens are as dull and sexually monochromatic as tropical wrens. Hummingbirds are generally sexually dichromatic, with brightly colored males and dull females, regardless of latitude. Swallows, vireos and flycatchers in either region are generally sexually monochromatic. There are always exceptions, such as the sexual dichromatism in the Inca Wren, *Thryothorus eisenmanni*, of Peru or the Vermilion Flycatcher, *Pyrocephalus rubinus*, of Central and South America. But where latitudinal differences occur, in groups like tanagers, warblers and orioles, temperate species tend to be dichromatic with dull females, whereas tropical species are more often monochromatic with both sexes brightly colored.

In blackbirds and tanagers, most of the differences among species in sexual dichromatism are the result of selection on female plumage color, rather than male plumage color (Irwin 1994, Burns 1998). Evolutionary changes in plumage are more common among females than males, and more often than not result in increased plumage ornamentation in females. There are at least three forces at play on female coloration:

1. female–female competition for territories and mates
2. female–male competition and
3. selection for dull coloration in females due to nest predation (Amundsen 2000).

Sexual monochromatism, where both sexes are bright, can indicate that strong sexual selection is acting on females as well as males. Female–female aggression, whether it occurs in defense of territories or in competition for access to males, can select for the same kinds of plumage signals we see in males. Jones and Hunter (1993, 1999) have shown that plumage ornamentation in the Crested Auklet, *Aethia cristatella*, affects mate choice and intrasexual aggression for both sexes. In this temperate species sex roles are close to equal, as for many tropical species.

Equal plumage ornamentation in males and females occurs in some lekking species, and is surprising because males experience such strong sexual selection (Stiles and Wolf 1979, Trail 1990). Sex roles are as divergent as one can get! Trail (1990) compared social interactions in the monochromatic Capuchinbird, *Perissocephalus tricolor*, with another cotinga which is sexually dichromatic, the Guianan Cock-of-the-Rock, *Rupicola rupicola* (Figure 6.7). Female Cock-of-the-Rocks are typical of many lek species because female aggression is rare, they visit several males before copulating, and assess males by passively observing their territorial and courtship display (Trail and Adams 1989). Capuchinbird leks are peculiar because of the very high frequency of chases and female–female aggression. Male dominance is determined by direct fights and chases, with the alpha male, who obtains all the copulations, chasing intruders away from his perch about once per minute during female visits. Female Capuchinbirds often arrive at the lek in groups, visit only the alpha male, and fight with each other for access to the alpha male's perch. Intersexual aggression is also common, with females being repeatedly chased by non-alpha males. Females sometimes give male-like displays during aggressive female–female interactions, and may also benefit from male-like plumage in intimidating males during chases. High female–female aggression, along with male–female chases, explains the convergence of sex signals in Capuchinbirds.

Female–female aggression in territory defense favors signals of dominance and fighting ability (e.g. Rohwer 1982). Year-round territory defense might be expected to favor bright female plumage, just as it favors female song. This appears true for the Red-winged Blackbird where females in Cuban populations defend territories year-round and are more brightly colored than females in northern populations (Whittingham *et al.* 1992b). But vigorous year-round territory defense is not always associated with bright female plumage. In the sexually dichromatic Dusky Antbird, females are strongly territorial towards other females (Morton and Derrickson 1996). Females do have the same white backspot as males, but are orange-brown instead of black. It is not known how important the black plumage is for male territory defense, relative to the white backspot. Similarly, there is no reason that orange-brown plumage is not as effective in signaling aggression among females as is black plumage for males. If aggression is primarily

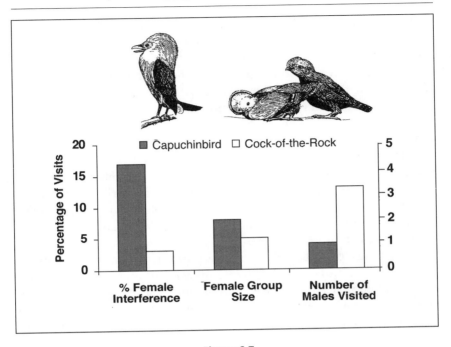

Figure 6.7

Comparison of female behavior at leks of the Capuchinbird and Guianan Cock-of-the-Rock (Trail 1990). Average female group size (at dawn, for the Capuchinbird), number of males visited per female, and percentage of courtship visits that ended in female–female displays or chases. Drawings from Johnsgard (1994).

intrasexual, females may use different but equally effective signals as males (e.g. West-Eberhardt 1983). Female Dusky Antbirds also sing in defense of territories, and their songs are also distinctive from the males.

Interspecific aggression, and male–female aggression could favor convergence in male and female signals (West-Eberhardt 1983). Many ant-following species are monochromatic, like the Bicolored Antbird, *Gymnopithys leucaspis*, Ocellated Antbird, *Phaenostictus mcleannani*. Intraspecific and interspecific aggression is certainly high when pairs from different territories converge at an ant swarm (Willis 1967, 1972, 1973, Willis and Oniki 1978), which may select for sex role convergence in plumage signals. The dominant Oscellated Antbird is also monochromatic, but in this species female aggression is very low and females literally cower behind their mates in intraspecific aggressive encounters (Willis 1973).

In tropical hummingbirds, bright female coloration evolves when both sexes compete for feeding territories (Wolf 1969, 1975, Bleiweiss 1985). Females defend their territories from either sex. Intersexual territoriality also occurs in migratory passerines that defend nonbreeding territories in the tropics. Breeding territories are defended against same-sexed intruders for several months in the temperate zone, but then each bird must defend a winter feeding territory alone for some six months against both sexes. In Hooded Warblers, *Wilsonia citrina*, females compete directly with males, as well as females, for winter territories (Morton *et al.* 1986, Stutchbury 1994). Females do not sing, but rather males and females both use call notes in territory defense, as noted above for many migrants. Plumage signals, like the partial black hood of female Hooded Warblers, may be important for territory defense in the nonbreeding season.

Male choice of mates can also favor plumage ornaments in females (Hill 1993, Amundsen *et al.* 1997, Hunt *et al.* 1999, Jones and Hunter 1999). When males invest heavily in parental care, a male's reproductive success may depend greatly on the quality of the female (reviewed in Cunningham and Birkhead 1998). Thus females may have to compete directly for high quality males, and use ornaments to advertise their quality to prospective males. In many tropical species males invest heavily in parental care because they build nests, incubate the eggs, and care for the young for extended periods. Species with bright female coloration certainly do not feature greater male investment in reproduction, but there is really nothing known about whether male choice of females is an important selective force on bright coloration in female tropical birds.

The other side of the color coin is the disadvantage of being conspicuous. As a general rule, predation favors inconspicuousness. In the temperate zone it is usually females that build nests and incubate, and this could favor dull coloration in females, and hence dichromatism, to reduce the likelihood of nest predation. Martin and Badyaev (1996) found a negative correlation between female brightness and frequency of nest predation in temperate zone warblers and finches. How does this apply to tropical birds? Nest predation is generally higher in the tropics (Chapter 3), but females tend to be brighter, not duller as one might expect. Bright female plumage is common in tropical warblers and tanagers with open cup nests as well as those with domed nests where birds are hidden from view while on the nest (Skutch 1985). Female plumage brightness can have little effect on risk of nest predation even for open cup nesting birds (e.g. Stutchbury and Howlett

1995). Brightness does not always equate with conspicuousness (Endler 1978), and even brightly colored birds can modify their conspicuousness by choosing habitats and lighting conditions that minimize their contrast with the background (Endler and Théry 1996).

It is frustrating not to find clear answers to the question of how selection pressures on plumage signals may vary between temperate and tropical regions. The solution is not more comparative studies searching for correlates of female coloration, but the application of experimental tools like model presentations and plumage dyeing. As with many topics in this book, our inability to find good answers rests largely with the lack of field experiments on tropical birds.

6.7 Immunocompetence and signaling mate quality

Parasites may be important to understanding temperate/tropical differences in the kinds of signals used in mate choice, and why they are used in assessment. Hamilton and Zuk (1982) first suggested the link between parasites, plumage brightness, and female mate choice. Folstad and Karter (1992) hypothesize that testosterone drives a tradeoff between plumage brightness and immunocompetence. Secondary sexual traits are costly for males to bear *because* they are testosterone-dependent. Elevated levels of testosterone suppress the immune system and increase a male's risk of mortality from parasites (viral, bacterial, protozoan, etc.). It is the underlying high testosterone which makes displays (vocal or visual) a handicap, and therefore a good way for females to assess male quality (Zahavi 1975, 1977, Zahavi and Zahavi 1997).

Males are apparently more susceptible to parasites than females (Poulin 1996, Møller *et al.* 1998a). This is consistent with the idea that males are vulnerable to parasites due to high testosterone. Immunosuppression has been linked to exaggerated male ornaments (Saino *et al.* 1997a, Verhulst *et al.* 1999) or displays (Soler *et al.* 1999) in birds. The role of testosterone in immunosuppression is not straightforward, however, since experimentally increasing testosterone does not necessarily reduce immune function (Hasselquist *et al.* 1999). Comparative tests (Møller 1997, Møller *et al.* 1998a) suggest that investment in immune function and immunosuppression increases with the intensity of sexual selection.

The immunocompetence hypothesis is clearly based on the ubiquitous temperate zone influence, and likely applies only to temperate zone birds. Testosterone levels that soar during the breeding season are the cornerstone of the idea (Folstad and Karter 1992). Many tropical birds have very low levels of testosterone year-round (Chapter 5; Wikelski *et al.* 1999a,b). Even lekking species, and those with EPFs, have low testosterone compared with temperate zone birds. Their low levels of testosterone presumably have little immunosuppressive effect. While Folstad and Karter (1992) suggest that any hormone or chemical could mediate the handicap, most studies to date have assumed testosterone is the culprit. Monogamous tropical species have obvious secondary sexual traits, vocal and visual, but these are not hormonal handicaps.

Secondary sexual characters could reliably reveal male immunity, regardless of testosterone levels, if energy expenditure to produce mating displays takes away from energy invested in immune defense (Deerenberg *et al.* 1997, Soler *et al.* 1999). In temperate zone birds high song output, or other displays, may be energetically draining enough to depress immune function during breeding, and thus function as a handicap. In many tropical birds sex roles are close to equal, with female song and territory defense being common. Immunosuppression, if it occurs at all, should occur fairly equally for both sexes during reproduction. We expect that immunosuppression will not be an important feature of mate assessment and signals for socially monogamous tropical birds.

7 | Biotic interactions

7.1 A whole new ecological stage for the evolutionary play

The first simple and broad generalization of this book was mentioned in the Introduction: species in temperate regions are under strong selection from abiotic factors (e.g. climate) whereas in tropical regions biotic selection pressures are most important. Interactions with other species (plant and animal) play a key role in shaping the behavioral adaptations of tropical birds. Examples were provided in earlier chapters. Breeding seasons and mating systems in the tropics often interact with fruit availability but rarely do so in the temperate zone. Some tropical territorial systems are defined by the biotic interactions that brought them about. Mixed species flock, army ant influenced, and fruit influenced (Table 5.1) are examples. But even territories used only for breeding are influenced by fruit availability in the tropics but not in the temperate zone (Morton 1973, Stutchbury et al. 1998). Predation may influence tropical birds more as evidenced by the stable territory boundaries found in tropical, but not temperate zone, 'year-longers' (Morton et al. 2000) and by nesting associations with ants and wasps (Janzen 1969, Robinson 1985, Wunderle and Pollock 1985, Joyce 1993). We will expand on this theme here.

Biotic interactions come in various forms but all imply an evolutionary influence of individuals of one organism or group of organisms on the individuals of another. Our focus in this chapter is not on obligatory mutualisms, such as between figs and fig wasps, nor on the looser forms of mutualisms lumped under 'coevolution,' whether or not they really are (McKey 1975, Janzen 1980, Davidar and Morton 1986, Davidar 1987, Reid 1987). Coevolution is an evolutionary change in a trait of the individuals of one population due to a trait of individuals of a second population, followed by an evolutionary response in the second population to the change in the first (Janzen 1980). Coevolution is a common occurrence throughout the world but its strict definition makes latitudinal comparisons of its impact difficult. Attempts to compare the tropics and temperate zones in terms of coevolution between, for example, birds and plants that produce fruit, is apt to be unproductive (Wheelwright 1988), even though the results of avian frugivory on plants differ dramatically (Morton 1973). Mistletoes occur at many latitudes but they influence entire genera of birds only in the tropics. Biotic interactions imply a more general and diffuse theme built upon selfish interactions and behaviors used by different organisms and birds. G. E. Hutchinson (1965) outlined the influence of ecology as a stage for the play of evolution and we feel this metaphor is valuable for understanding latitudinal differences in the importance of biotic interactions in shaping evolution.

7.2   Birds and plants

David Snow's adage that 'fruit wants to be eaten while insects don't' underpins some general worldwide trends that seem to vary little with latitude (Snow 1971). Plants compete to have their fruit eaten, producing staggered fruiting seasons (Snow 1965), conspicuous colors and displays of fruit, and fruit of a size that is easily swallowed. Insects, on the other hand, hide so as to avoid detection and predation by birds. This predation pressure results in a multitude of ways and places in which insects hide which, in turn, provides different foraging niches for the birds pursuing them.

In the temperate zone few plants produce fruit during the avian breeding season (Figure 7.1). The fact that plants do *not* ripen fruit during the temperate zone spring when birds are breeding results from the birds' lack of interest in fruit then. Many wind-dispersed trees, such as red maple and elms, ripen seeds during or before they leaf out in the spring. Thus, plants are physiologically capable of fruiting early in the spring but they do not produce bird-dispersed fleshy fruit in the temperate zone spring and early summer because there are too many insects! Temperate zone birds have an abundance of insects and rely heavily on caterpillar larvae to feed themselves and their young (e.g. Holmes and Schultz 1988, Naef-Daenzer and Keller 1999) and do not have the time and energy budget options that fruit provides for tropical species. Moreover, breeding temperate zone birds are sedentary and more restricted to their territories than tropical birds. They would make poor seed dispersers on that account as well.

As a result of these factors, plants rarely fruit during the avian breeding season in temperate latitudes (Figure 7.1). Even though 'fruit wants to be eaten and insects don't,' the sheer abundance of insects,

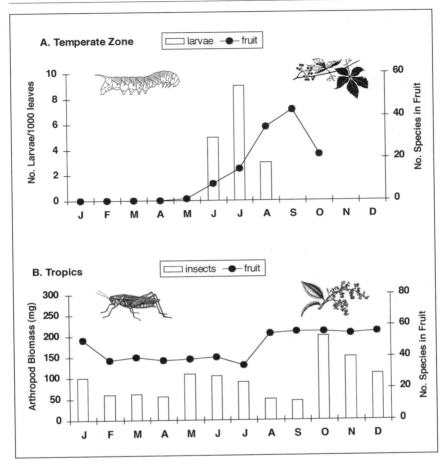

Figure 7.1

Insect and fruit abundance over the year in the temperate zone versus tropics. Caterpillar data from New Hampshire (Holmes *et al.* 1986) and arthropod data (biomass per 30 sweeps) from Costa Rica (Young 1994). Fruiting data from Illinois (Thompson and Willson 1979) and Costa Rica (Levey 1988).

particularly caterpillars specialized to feed upon newly emerging leaves, makes fruit less appealing to birds. The avian altricial nestling growth strategy requires high protein foods for the rapid growth of nestlings. Such proteins are too costly for plants to invest in attracting them to eat fruit (Snow 1971). They lose out in the insect-rich temperate zone and shift production. Temperate zone plants produce fruit primarily after birds breed. The reasons are obvious when viewed from the plants' perspective. Breeding territories are no longer defended, allowing birds to search for and concentrate at fruiting plants. Bird populations are highest just after the birds have fledged their young. Birds now choose fruit, and plants have responded by producing it after the breeding season for temperate zone birds (Thompson and Willson 1979, Stiles 1980, Willson 1983, Fuentes 1992). Perhaps the most comprehensive study of temperate zone bird–plant interactions is the delightful book by Barbara and David Snow (1988).

Fruit is often so abundant in the tropics that there is little competition for it. For example, Willis (1966) observed 28 species of birds feeding on *Conostegia* berries in Colombia. Nine of these species were of one tanager genus, *Tangara*. This is a spectacular sight, and is especially common around 1500–2500 m elevation in the neotropics. Fruit eating promotes more fruit and the living is easy. The ways in which birds eat fruit is an important source of selection on plants but is beyond the scope of this book. A vast and excellent literature on bird–plant interactions exists (e.g. Herrera 1981, Snow 1981, Howe and Smallwood 1982, Davidar 1983, Estrada and Fleming 1985, Moermond and Denslow 1985). Levey *et al.* (1994) is highly recommended for an overview and example of the importance of avian frugivory in the La Selva forest of Costa Rica.

Over evolutionary time it might be expected that totally frugivorous diets in birds would increase. The abundance of fruit, the short time needed to obtain it, and its year-long availability in the tropics (Snow 1965) would seem to provide an adequate and welcoming food supply for adults and young. Even a grass, *Lasiacis* spp., common in forest edges, has evolved a seed that so resembles a black fleshy fruit that only frugivorous birds eat it, not grain eaters (Davidse and Morton 1973). A fruit diet for nestlings would provide food for larger clutches than hard-to-find insect food. But, despite that, fruit is not generally used as food for small young (Lack 1968).

When both adults and nestlings are sustained exclusively with fruit we call the species a total frugivore. But they are rare in the tropics and nonexistent in the temperate zone (Figure 7.2). Even species generally considered frugivorous, such as most manakins (Pipridae) and many cotingas (Cotingidae), are not *totally* frugivorous, since they feed their young partially or wholly on insects (Skutch 1969). We call these 'adult frugivores,' when adults eat fruit but still feed their small nestlings invertebrates (Figure 7.2). Little is known about the diet of adult birds when they are feeding nestlings. More information on this question would be valuable because the foods eaten by adults while they are feeding nestlings has important consequences for breeding time-energy budgets.

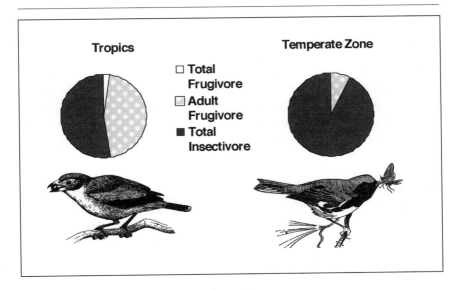

Figure 7.2

Proportion of passerine genera that are total frugivores (adults and nestlings eat fruit), adult frugivores (adults eat fruit, nestlings eat insects), and total insectivores (adults and nestlings eat insects) in Panama (n = 204 genera) and North America (n = 94 genera). Drawings of temperate Black-throated Blue Warbler from Griscom and Sprunt (1957) and tropical Masked Tityra from Wetmore (1972).

There are four ways in which diets could be divided between adults and their nestlings (excluding granivory): 'total frugivory' where adults eat fruit and feed fruit to nestlings, 'total insectivory' when adults eat insects and feed insects to nestlings, 'adult frugivory' when adults eat mostly fruit and feed insects (and fruit) to nestlings and, a final category, 'nestling-only frugivory' in which adults eat insects but feed fruit to nestlings. Most temperate zone passerines are total insectivores, and adult frugivory is very common in the tropics. There are no species in the entire world in the 'nestling-only frugivory' category! Many temperate zone birds, such as Rose-breasted Grosbeaks, *Pheucticus ludovicianus*, Scarlet Tanagers, *Piranga olivacea*, Red-eyed Vireos, *Vireo olivacea*, and Gray Catbirds, *Dumetella carolinensis*, would certainly be in the 'adult frugivory' club if fruit was available for them during the temperate zone breeding season. There are exceptional plants too. In eastern North America these are bushes and small understory trees such as juneberry, *Amelanchier* spp., and red elderberry, *Sambucus racemosa*, which do fruit in the latter half of the breeding seasons of these birds and are relished by them.

Fruit is conspicuous and is easily and quickly obtained. As a result, adults eating fruit have more time to find nestling food. By not eating invertebrates that might be fed to nestlings, they are not competing with their own young for this food. It is predictable, then, that most adult tropical birds do indeed feed invertebrates to their nestlings but feed themselves on fruit. This can best be seen in relation to territorial systems (Table 5.1). Species in the 'breeding territory' category include all the granivores (e.g. *Sporophila, Oryzoborus, Jacarina, Tiaris*) but also many adult frugivore species such as Yellow-green Vireo, *Vireo (olivaceus) flavoviridis* Lesser Elaenia, *Elaenia chiriquensis*, and Piratic Flycatcher, *Legatus leucophaius*, which are intra-tropical migrants (Morton 1977a). All lekking and all fruit-influenced territorial species are adult frugivores or total frugivores, as are many of the year-long territorial species (e.g. *Mimus*, all the vireonids, *Icterus, Atlapetes, Arremonops, Saltator, Chlorothraupus, Habia*, etc.).

Total frugivory would be common if time/energy budgets for breeding were the only concern, but they are not. A drawback to fruit-eating is that it undermines the ability of their nestlings to grow extraordinarily fast, so fast that it is called the 'altricial nestling strategy.' Altricial nestling birds, as opposed to precocial ones like chicks and ducklings, are poikilothermic (cold-blooded) for the first several days after hatching. Adults brood to maintain the body temperature of their altricial young at or near their own temperature. Energy from nestling food is fully used by them to grow rather than to regulate temperature (Dawson and Evans 1957). It is this metabolic saving that is called the 'altricial strategy,' and it results in nestlings fledging far faster than would otherwise be possible.

The total frugivore strategy is rare because fruit does not contain sufficient protein to permit rapid growth and development (Morton 1973). The species that do fit this category can enlarge clutch sizes far beyond that standard in the tropics, if they have a safe nesting site (Morton 1973, Sargent 1993), for their parents can gather an unlimited amount of food. However, the time young spend in the nest before fledging increases as the amount of fruit in the diet increases (Skutch 1954, 1960). Comparative data show this (Figure 7.3). The same is true for non-oscine passerines (Figure 7.3, Skutch 1969). B. K. Snow (1970) found that a comparable, totally frugivorous non-oscine, the Bearded Bellbird, *Procnias averano*, has a lengthy nestling period lasting 33 days.

The altricial nestling strategy is an adaptation for shortening the vulnerable period in the nest. This strategy has been particularly valuable

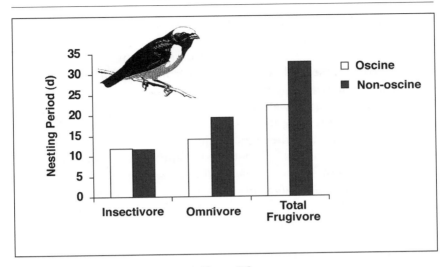

Figure 7.3

Average duration of nestling period for total insectivores, adult frugivores and total frugivores (like the Yellow-crowned Euphonia shown), for oscines and non-oscines. Data from Skutch (1954, 1960, 1969) and Snow (1970). Drawing from Skutch (1954).

to passerine birds in allowing them to nest in the 'open' whereas the nonpasserines, with exceptions, nest in holes affording greater safety. This gives passerines a flexibility and freedom to nest almost anywhere (Moynihan 1998). A fruit diet for nestlings cannot sustain the altricial growth potential with the result that predation pressure will be greater.

In many tropical passerine birds, 75% or more of nests fail to produce young owing to predation (Ricklefs 1969b). In observing a 75% mortality in a sample of 100 nests, we are left with 25 nests at the end of a 25-day nesting cycle, which includes incubation and nestling stages. If this nesting cycle were lengthened by 9 days by frugivory, another 9.8 nests would be lost to predation, leaving 15.2 successful nests instead of 25. Dusky Antbirds, *Cercomacra tyrannina*, may be typical. Their nests, even with only a 9-day nestling period, have a mere 8% chance of producing independent young (Morton and Stutchbury 2000). Predation imposes a cost on feeding fruit to nestlings and is the reason why birds do not nourish them with such an abundant food, at least while they are little and cold-blooded.

But, surely parents would feed fruit to their nestlings if they were starving! Observations suggest that this is not the case. For example, nestling starvation was common in Clay-colored Robins, *Turdus grayi* (Chapter 2) even though *Panax* and *Miconia* berries, figs and papaya, which were fed to nestlings occasionally, were present in abundance. By experiment, we found that there is an upper limit to the amount of fruit nestlings will accept, even when hungry. Captive robin nestlings (2 days old), after several hours on an all-fruit diet (*Bursera simaruba, Miconia argentea*, papaya, banana, figs, palm fruit), still gaped hungrily for food but refused to swallow more fruit. Crickets, mealworms, and ground meat were readily accepted by these captives (Morton 1973).

This suggests that, in the wild, nestlings exercise some control over their diet. Species may vary widely and adaptively in this trait. Two to four day-old Yellow-bellied, *Elaenia flavogaster*, and Lesser Elaenias were fed a mix of fruit (*Panax morototoni*), tree hoppers, stingless bees, and some orange spiders about 2.3 mm long. After day six, the nestlings were fed almost exclusively on fruit (Morton *et al.* unpubl.). The highly frugivorous Northern Mockingbird, *Mimus polyglottus*, does not feed any fruit to nestlings until they have reached the age of 6 days and have begun thermoregulating (Breitwisch *et al.* 1984). There is great variation in the timing and amount of fruit fed to nestlings and almost nothing is known about it. Ecologically important adjustments to local foraging conditions might occur if adults vary the proportions of fruit and insects they feed to nestlings over time. That the tropical fruit/insect strategy is so ubiquitous attests to its success as a biotic adaptation.

Comparisons of closely related birds nicely illustrate this temperate/tropical difference in bird–plant interaction. Contrast the temperate American Robin, *Turdus migratorius*, with the tropical Clay-colored Robin. They are closely related, highly frugivorous as adults, similar in size, use gardens and lawns for foraging, and have extra-pair mating systems. American Robins, during their breeding territoriality, are restricted to invertebrate foods, largely earthworms, and have an insect/insect adult and nestling diet. After breeding and throughout the nonbreeding period of the year, they become almost total frugivores (Wheelwright 1986, Levey and Karasov 1992). They switch to invertebrate food during early spring when most overwintering fruit has been consumed or rotted. When it is cold and earthworms are too deep to reach, they consume insects that have overwintered as caterpillars, some of which (e.g. woollybears, *Isia isabella*) need a great deal of scraping to remove setae before they can be swallowed (Morton 1968). Clay-colored Robins also have the breeding territorial system but leave their territories to feed on fruit. Males and females probably spend only half of their time on territory rather than nearly all of it as in the American Robin. Female Clay-colored Robins will travel across a

dozen territories to reach a favorable foraging area to fetch invertebrates for their nestlings. Clay-colored robins use a special vocalization
that sounds like *skeetch* when they are trespassing *en route* to fruit
sources; the high pitch suggests that it is appeasing to the territory
owner. American Robins have a similar sibilant note that is used as a
'flight call' by birds in frugivorous fall flocks. These thrushes differ only
in the importance fruit has for them while *breeding*. Their difference
reflects the dissimilar tropical/temperate zone pattern described above.

Latitudinal trends in bird–plant interactions are more complicated
than viewed earlier. It was suggested that plants evolved more nutritious fruits as the dispersal quality of the avian frugivores increased
(Snow 1971, McKey 1975, Howe and Estabrook 1977), a true coevolution with equal tradeoffs between costs and benefits in the birds and
plants. In reality, the idea that birds and plants are coevolved was a
logical suggestion but has not been verified (Herrera 1981). Neither
nutrition of fruits nor the size of the crop and its display seem coevolved
with avian dispersers (Davidar and Morton 1986). The reason for this
is that a third party is involved. It is not simply birds and plants but
predators, birds, and plants that interact (Morton 1973, Howe 1979).
Thus, while avian frugivory has had a huge effect upon the world's
ecology, perhaps because the angiosperms are reproductively superior
to gymnosperms owing to their production of fleshy fruit (Regal 1976),
this effect is likely due to more general biotic interactions than implied
by the term coevolution (Midgley and Bond 1991).

However, plants should do everything they can to lure birds to eat
their fruit and to disperse their seeds. If a plant can produce a laxative
to increase passage rate of seeds to its advantage (Murray *et al.* 1994),
why not addiction too? Plants have evolved many chemicals to protect
their fruit against rot (Cipollini and Stiles 1993a,b) and birds have
adapted behaviorally (Foster 1987a) and morphologically to a diet of
fruit (Richardson and Wooller 1988). So far, although mentioned in the
popular literature (J. Greenberg 1983), no one has documented a plant
that addicts birds to its fruit, even though this would appear possible.

7.3 Biotic interactions and latitudinal adaptations in migratory birds

Migratory birds, like the two *Turdus* thrushes, offer examples of
adaptations to tropical and temperate regions. Here we have an opportunity to compare the same individuals. Migrant birds adapt to all the
regions they inhabit and offer examples of latitudinal changes in the

importance of biotic interactions (Rappole *et al.* 1983, Rabøl 1987). Even though it was generally thought that most of these migrants leave the temperate winter because their insect food disappears, nearly all species become partly or highly or even totally frugivorous during the tropical portion of their annual cycle. As a whole, the migrant group is indistinguishable from resident tropical birds in their reliance upon plants for fruit or nectar while they reside in the tropics (Rappole 1995) and during migration as well (Parrish 2000). The physiological changes that accompany this diet switch is a fascinating story in its own right (Levey and Karasov 1989, Martinez del Rio and Karasov 1990). Even within the tropics migration is a search for fruit, not insects. All intra-tropical and elevational migrants travel to find fruit (Levey and Stiles 1992). Yellow-green Vireos and Piratic Flycatchers, for example, leave breeding territories in Panama in the wet season for dry areas in South America where fruit is in greater abundance (Morton 1977a). Migration within and to the tropics, in other words, is based upon the biotic interactions of plants and birds.

This discovery came as a great shock to some temperate zone community ecologists, who assumed food habits were consistent during the year. For example it was mysterious why Cape May Warblers, *Dendroica tigrina*, have fuzzy-tipped tongues, of the sort used to harvest nectar. They eat insects in the temperate zone. Sure enough, this morphological attribute, used rarely in North America (Sealy 1989), fits the warbler's tropical habits well (Staicer 1992). In Cuba, for example, Cape Mays are eager nectar and fruit eaters. They are also quite pugnacious inter- and intra-specifically. One bird, traveling peacefully with a flock of resident *Terretristis* warblers, chased its larger flock mates away when they began pecking at a 5 cm-long red-ripe cactus fruit. Tennessee Warblers, *Vermivora peregrina*, gregarious and social when feeding on insects, become antagonists when feeding on the nectar of *Combretum*. *Combretum* has bright red pollen, which soon coats the heads of warblers tough enough to defend small patches of this sought-after plant. Thus 'war painted,' Tennessee Warblers take pollen to new *Combretum* vines, the red pollen paving the way by marking the dominant birds.

Similarly, a large tree, *Erythrina fusca*, attracts Orchard Orioles, *Icterus spurius*, to its large leguminous flowers. Orchard Orioles open the flower 'correctly' by scissoring open the large flag petal, exposing the nectary and getting pollen on their foreheads in the process. Once thus opened, a burnt orange color is displayed in the petals surrounding the nectary, the same color as that of the dominant male Orchard

Oriole. It is possible that trees with mostly opened flowers signal to Orchard Orioles that they are occupied and that it would be best for the birds to find other *Erythrina* trees, thus dispersing pollen (Morton 1979b).

The Eastern Kingbird, *Tyrannus tyrannus*, is frugivorous, preferring the nutritious fruit of *Sassafras albidum* when leaving its nearctic breeding areas and *Panax morototoni* throughout its tropical range (Morton 1971a). In fact, this kingbird migrates first to the far southern end of its nonbreeding range and then moves northwards with the dry season, apparently tracking the ripening of *Panax*.

A last example is the red male coloration in Scarlet, *Piranga olivacea*, and Summer Tanagers, *Piranga rubra*. As the scientific name suggests, Scarlet Tanager males turn olive-green in August before migrating to South America whereas the male Summer Tanagers retain their red plumage all year. Scarlet Tanagers are, arguably, even redder than Summer Tanagers, with a rich scarlet tone as opposed to the Summer Tanagers' comparatively dull redness. All the carotinoids going into these bright colors are derived from the birds' diets. We think that the more brilliant red of the Scarlet Tanager is derived from its habit of consuming the bright red arils found around the seeds of *Tetracera* spp. vines. These vines, and, in fact, arils in general, are restricted to the tropics. Arils are often very nutritious with lipid and protein components reaching 60 and 15 percentage of dry weight of the fruits (Foster and McDiarmid 1983). The Summer Tanager may be restricted to a dull red by its need to find carotinoids during its temperate zone molt, which are unlikely to match the red-orange concentrate of *Tetracera* arils. When courting a female, a Scarlet Tanager male positions himself below her, droops his black wings, and displays his incredibly scarlet back. This should be regarded as a tropical biotic interaction that has been carried north to continue operating on a temperate breeding territory. Such ties to tropical regions underscore the importance of tropical adaptations to the vast hordes of birds that winter there.

Plants often seem to specialize on migrant birds, perhaps because they move seeds longer distances than do residents. To attract migrants, leaves of many of these plants turn red early (e.g. Virginia-Creeper, *Parthenocissus* spp.) or have pale or rusty leaf undersides which flash in the tropical dry season tradewinds (e.g. *Cecropea, Panax, Miconia*). The term 'fruit flags' has been applied to describe this adaptation for conspicuousness to birds migrating above (E. W. Stiles 1980).

In other cases, migrants eat more fruit than closely related residents. In the Yucatan Peninsula of Mexico, White-eyed Vireos, *Vireo griseus*, migrants from North America, defend individual territories that overlap extensively with those of the resident Mangrove Vireo, *Vireo pallens*. The White-eyed Vireo requires *Bursera simaruba* trees, whose fruit ripens slowly over the stay of the vireo from September to April (Greenberg *et al.* 1993). It could not overwinter in the Yucatan if it were not for this one tree species (Greenberg *et al.* 1995). Greenberg (1981) also found that the red arils of a small tree found on Barro Colorado Island in Panama, *Lindackeria laurina*, were only eaten by overwintering wood warblers (Parulinae). The gregariousness of these birds may make them important dispersal agents. Such close associations between nearctic migrant birds and certain tropical plants may be fairly common.

Indeed, the tropical nonbreeding ranges of many migrant species, or the overwintering movements of the birds, may be directly related to locality-specific fruiting patterns (Morton 1980). In central Panama, Bay-breasted, *Dendroica castanea*, and Tennessee Warblers moved from forests during the wet season half of their stay (Oct.–Dec.) to forest edge where *Miconia argentea* was a favorite fruit. There they were joined by Chestnut-sided Warblers, *Dendroica pensylvanica*, which abandoned their forest territories for the fruit (Greenberg 1984a). All species underwent extensive and rapid molt of body feathers only after the fruit became available. Because this fruit ripens in the dry season, and the timing of the dry season in Panama accommodates their molt before their northward migration, the winter ranges of these warblers may be influenced by this calendar of seasonal effects on fruit.

While they are in their neotropical ranges, many migrants join mixed species flocks. There is no suggestion that migrants avoid resident tropical species that consume foods similar to their own (Table 7.1). Indeed, the mixed species flocks joined by the most common overwintering warblers and vireos in Panama are found in the canopy, and these contain the birds most similar to them. Chestnut-sided Warblers, Bay-breasted Warblers, Tennessee Warblers, Philadelphia, *Vireo philadelphicus*, and Yellow-throated Vireos, *Vireo flavifrons*, are attracted to flocks led by Lesser Greenlets, *Hylophilus minor*, the tropical equivalent of the temperate zone chickadees in their attractiveness to flock participants (Morton 1980). Migrant birds thus are important attendants of tropical mixed species flocks, groups of birds with far more complexity than found in their temperate zone counterparts.

Table 7.1

Resident species that occur in mixed species flocks in Panama that are joined by North American migrants (from Morton 1980). Habitat indicates forest canopy, forest edge or forest understory.

Species	Habitat	Mass (g)
Plain Xenops	Understory	12.0
Checker-throated Antwren	Understory	10.4
White-flanked Antwren	Understory	8.5
Dot-winged Antwren	Understory	11.0
Ruddy-tailed Flycatcher	Understory	7.4
Yellow-margined Flycatcher	Understory	14.2
Southern Bentbill	Understory	6.9
Yellow-green Tyrannulet	Canopy	7.0
Forest Elaenia	Canopy	12.0
Southern Beardless Tyrannulet	Canopy, Edge	6.0
Paltry Tyrannulet	Canopy, Edge	7.2
Brown-capped Tyrannulet	Canopy, Edge	6.5
Tropical Gnatcatcher	Canopy	6.7
Lesser Greenlet	Canopy	9.3
Shining Honeycreeper[a]	Canopy, Edge	12.2
Red-legged Honeycreeper[a]	Canopy, Edge	12.0
Green Honeycreeper[a]	Canopy, Edge	13.0
Blue Dacnis[a]	Canopy, Edge	10.0
Fulvous-vented Euphonia[a]	Canopy, Edge	11.5
Golden-hooded Tanager[a]	Canopy, Edge	20.5
White-shouldered Tanager	Canopy	13.0

a: Fruit or nectar eating species, joined mainly in dry season.

7.4 Avoiding predators

How some cope

Predator avoidance by birds is nearly as diverse as that shown by insects trying to escape predation by birds. Perhaps it is equally diverse, now that poisonous birds in the genus *Pitohui* have been added to the list of birds suggested to be merely distasteful to predators (Cott 1940). Pitohuis live in New Guinea yet share a toxin, batrachotoxin, that is also found in poison dart-frogs (*Phyllobates*) found in the neotropics (Dumbacker *et al.* 1992).

For the majority of tasty birds, however, hiding is a better option. Studies on lekking birds in the tropics have shown a remarkable degree of association between plumage color patterns and the exact location and timing of displays (Endler and Théry 1996). This association results from the ever-present threat of predation. By displaying in certain light environments, birds can be conspicuous to each other but not to predators. Let's look in some detail at this example.

Cock-of-the-Rocks display most often mid-morning and mid-after-noon, and only when the sun is not blocked by clouds (Figure 7.4A). White-throated Manakins, *Corapipo gutturalis*, display most often at mid-day, and usually when the sun is out (Figure 7.4B). The White-fronted Manakins, *Lepidothrix serena*, displays most often very early and late in the day, when the sun is still below the horizon on the mountain sides where leks occur (Figure 7.4C). When the sun is above the horizon, males display when clouds block the sun. How do we make sense of the differences in display behavior between these species? Are they arbitrary results of sexual selection? No, they reflect behavioral adaptations to minimize the risk of predation.

The Guianan Cock-of-the-Rock, *Rupicola rupicola*, is one the most spectacularly colored neotropical birds, with bright orange body and white wing strings. Each male defends a small display site consisting of a horizontal branch about 2 m above the forest floor which is cleared of leaves below (Endler and Théry 1996). Males display when part of their body is illuminated by a sun speck, and the rest of the body is in shade. Displays cease when clouds obscure the sun, or when the sun speck moves off the display site. This display behavior is not arbitrary.

All patches of the plumage pattern (rump and back, strings, light wing bar) show their maximum reflectance at longer wavelengths. The forest shade environment where leks are located are rich in these iden-tical wavelengths. The average overall brightness of the male is least conspicuous against the visual background during displays, when males are partially sunlit. A predator viewing the displaying male from a distance, without discerning individual plumage patches, would have difficulty picking out the bird against the background. At close range, the contrast in brightness and hue among the different plumage patches, and the contrast between the bird and the background, is max-imized by males displaying partly illuminated by the sun.

The display behavior of the two manakin species also maximizes their contrast with the background light environment during displays. Male White-throated Manakins display at the edge of a sun patch, with their white throat and chest in the sun and their dark back and head in the shade. They display on a log, and jump between opposite sun patches when a female arrives. Male White-fronted Manakins have a turquoise rump patch which is maximally visible in the dark light

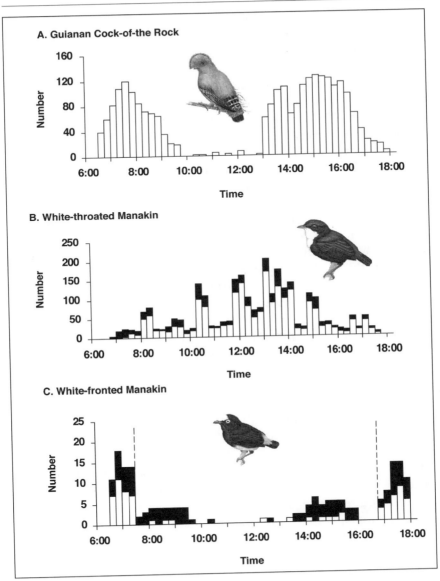

A. Guianan Cock-of-the Rock

B. White-throated Manakin

C. White-fronted Manakin

Figure 7.4

A) Timing of male lek displays versus time of day in A) Guianan Cock-of-the-Rock B) White-throated Manakin and C) White-fronted Manakin. Each graph shows number of observations versus time of day (white bars are sunny conditions, dark bars are cloudy conditions). For White-fronted Manakin (C) display behavior differs when sun is below the horizon (before 7:30 and after 3:45) compared with when sun is above the horizon. Figures modified from Endler and Théry (1996). Drawings from de Schauensee and Phelps (1978).

conditions used during display, which also minimizes the visibility of other body regions.

The color patterns and behavior of birds represents a tradeoff between crypsis to predators and conspicuousness to conspecifics. The habitat for lek locations, the times of day to display, and the precise display behavior are all molded by the constraints imposed by the light environment. While most easily studied in lekking species where displays occur at specific sites, these principles apply to all species.

This research has important implications for the conservation of tropical birds. Display sites are not arbitrary locations, as they might otherwise seem. The lighting conditions are rather precise, and even small disturbances on the forest can greatly change the light properties. A slight disturbance, due to trail construction or selective logging, can result in abandonment of traditional lek sites (Endler and Théry 1996). It is unknown how important particular light regimes are to territorial species, as this has never been studied, but one aspect of habitat suitability could include very subtle (to us) lighting conditions that are crucial for predator avoidance.

How most cope with predators – Mixed Species Flocks

'Wherever one travels on this earth, birds gravitate into mixed foraging parties. This is true from the shores of the Arctic Ocean to the equator, in humid forests as well as shrub deserts. The selective forces that promote aggregative behavior are thus virtually universal and independent of climate or vegetation' (Terborgh 1990). Predation on birds by hawks is commonplace around the world and one common countermeasure taken by prey species is flocking (Bates 1863, Morse 1970, 1977). In forests, such flocks consist of many species and are termed mixed species flocks. In open habitats, flocks tend to be monospecific and often quite large (Morton 1979a). Here, we will not focus on the anti-predatory nature of flocking but upon its social and ecological consequences. There are, however, two predatory aspects that deserve mention in this context. While many eyes are able to detect predators, so-called 'alarm notes' given by flock members may actually benefit the caller more than the perceivers (Charnov and Krebs 1975).

Another predation-related aspect is how absorbedly the birds forage. Tropical birds are generally slow foragers that search intently for cryptic or hidden prey, compared with temperate zone birds that rely heavily on caterpillar prey (Figure 7.5). Temperate species most often glean prey from foliage (Figure 7.5) and forage at a high rate (Thiollay 1988). Tropical passerines feed in wider variety of substrates, using a wider variety of techniques for prey capture. This basic difference in food availability and foraging styles has a big impact on social behavior. Birds that forage intently, for instance by peering inside aerial dead leaves, can forage best if in company with less intent foragers, such as those that scan long distances for flying insects (Willis 1972). Which came first, flocking or intent foraging, is a valid question, but it is true that dead leaf foragers and others that cannot scan for predators and forage at the same time are often restricted to mixed species flocks. This

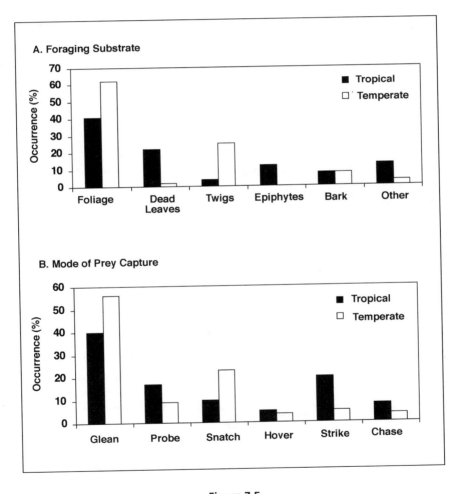

Figure 7.5

Comparison of the overall frequency of occurrence of different foraging substrates and modes of prey capture for tropical passerines in French Guiana and temperate passerines in France (after Thiollay 1988).

includes some nearctic warblers such as Worm-eating, *Helmitheros vermivorus*, Blue-winged, *Vermivora pinus*, Golden-winged, *V. chrysoptera*, and Black and White, *Mniotilta varia*, that join mixed species flocks of tropical birds consistently if not obligatorily.

While mixed species flocks are found worldwide, like so many other aspects of avian adaptation, mixed species flocks in the tropics are far richer in biotic interactions than those in the temperate zone. One reason for the difference is that flocks are primarily a nonbreeding phenomenon in temperate latitudes. There are no temperate species that breed in mixed species flocks (Table 5.1). Temperate flocks break down when their members disperse to breeding territories. Nonetheless, being in a flock has important consequences even in simple temperate zone nonbreeding flocks. Flocking can promote efficient foraging because birds in flocks spend less time on the lookout for predators than birds alone (Sullivan 1984). But aggression and social dominance have much to do with where a species typically forages (Morse 1974) as well as where the individuals within a species feed (Schneider 1984). There may also be gender differences in foraging such as occurs in the Downy Woodpecker, *Picoides pubescens* (Peters and Grubb 1983).

In contrast, tropical mixed species flocks are a dominant feature in neotropical forests throughout the year (Munn 1985, Powell 1985). But, although many tropical birds have year-long territoriality (Chapter 5), those in permanent mixed species flocks are special in that regard. Their territorial boundaries may overlap to form a shared boundary for all the regular members of the flock. In Amazonia, mixed species flocks of the forest understory consist of pairs of 10–20 species. At least half of these species maintain territories that correspond exactly to the home range of the entire flock. Powell (1979) notes that permanent members vary in size, with smaller species occurring at the same density as larger ones. Single pairs of four small species (8 g body mass) occupied exclusive territories of 8–12 ha, the same area as occupied by six larger species (~37 g). Powell suggests that because the home range is determined by the needs of larger birds, and because the smaller species exclude conspecifics from the group home range, the smaller birds must under-utilize the food resources available to them. This under-utilization, in turn, predicts that smaller species can coexist with greater niche overlap, resulting in more species diversity of small species. He tested this idea by comparing the foraging overlap of small and large species. Sure enough, smaller species had greater foraging overlap, particularly the three smallest species in the genus *Myrmotherula*. *Myrmotherula* species do differ, however. In Peru, Munn and

Terborgh (1979) showed that Plain-throated Antwrens, *Myrmotherula hauxwelli*, and Pygmy Antwrens, *M. brachyura*, foraged very low and very high, respectively. White-eyed Antwrens, *M. leucophthalma*, and Ihering's Antwrens, *M. iheringi*, specialized on dead leaves and the undersides of vine stems and dead twigs, respectively. Gray Antwrens, *M. menetriesii*, foraged higher than the White-flanked, *M. axillaris*, and Long-winged Antwrens, *M. longipennis*, which were similar in foraging height, techniques, and where they looked.

The frugivore/omnivores tend to be less stable in terms of the individuals comprising the flocks and they are found mainly in forest canopy and edge. Many of these have fruit influenced territories and leave flocks while breeding. Moynihan (1962) studied the 'blue and green tanager and honeycreeper' flocks, and answered the question of how the flocks' members stay together. Some species actively follow others while others are simply followed. Some species contribute to flock formation, as either followers or those that are followed, while others do much less to stimulate the formation or maintain the cohesion of mixed flocks. Moynihan called species that contribute to flock formation 'nuclear' and others 'attendant.' Others describe 'core' species, those which occur in all understory flocks and share the jointly held flock territory. They describe other species with smaller, close-packed territories, that join a flock when it enters their territory, species that opportunistically join both canopy and understory flocks and, lastly, species with patchy foraging habitat that follow flocks regularly but only as long as they remain within appropriate habitat (Munn and Terborgh 1979).

The flocks composed of insect-consuming birds are the best studied. With the exception of ant-following birds, insects are dispersed and their predators must often work intently to find them or extract them from hiding places. Predation pressure on insects results in a multitude of ways and places in which they hide which, in turn, provides different foraging niches for the birds pursuing them. Thus we have bird species that glean bark, some that usually sally out from perches, and others peck inside dead leaves, etc. These birds often occur together and the different ways and places by which they exploit insects reduces competition and allows their coexistence in mixed species flocks.

Is this competition scenario true? On the depauperate avifauna of Cocos Island, 500 km southwest of Costa Rica where only four landbird species exist, individual Cocos Finches, *Pinaroloxias inornata*, use specialized behaviors to exploit insects. Considering the population of finches as a whole, their range of feeding behaviors span those of several families of birds on the mainland. There are no morphological differences underlying the foraging specializations, which ranged from probing or gleaning from leaves, branches or dead leaf clusters to taking nectar from flowers or extra-floral nectaries to gleaning insects or seeds from the ground. Instead, each bird may learn the behaviors associated with its foraging specialty (Werner and Sherry 1987). The foraging diversity of these Cocos Finches suggests that foraging specialization on a species level arises when there are many species competing for invertebrate food. Individuals of each species must be better than anyone else at procuring food in a precise manner. The behavior and the morphological tools underlying these special skills seem to be based on genetic differences between species (Greenberg 1983, 1984b, 1987). But the selection pressure favoring feeding divergence and specialization

Table 7.2

Geographic variation in the average flock size and species richness (total number of species present, and number of species regularly present) of insectivorous mixed species flocks in humid low-elevation, humid middle-elevation and dry tropical forests (from Powell 1985).

Location	Nucleus Species[a]	Flock size	Number Species	Number Regulars
Humid-low				
Amazonia	F	30–35	48	16
Amazonia	F	25–30	35	14
Venezuela	V	–	42	7
Panama	F	6	22	5
Panama	F	–	28	7
Panama	F	8	40	7
Panama	F	7	34	8
Mexico	F–V	–	44	–
Honduras	F–V	10–15	67	3
Southern Brazil	Th	–	20	6
Costa Rica	F	–	31	8
Humid-middle				
Panama	Th	8–15	21	8
Costa Rica	P	8	43	5
Colombia	F	22	46	10
Dry				
Mexico	T	40	10	3
Brazil	P	–	10	5

a: F, Formicariidae; T, Troglodytidae; V, Vireonida; P, Parulinae; Th, Thraupinae.

probably results from intense territorial behavior. Only those individuals that can exclude conspecifics can become the sole member of the pair of its species within a given flock, and reproduce. The efficiency with which that individual can obtain food, relative to other conspecifics, probably has to do with its ability to extract and retain prey in a mixed species assembly (Graves and Gotelli 1993).

As a result of this potential for foraging specialization on invertebrate prey, and the addition of omnivores, tropical mixed species flocks can contain many species. Understory flocks in Peru can contain about 42 species. Total species ranges from 10 to 67 (Powell 1985, Table 7.2). There have been some reported disadvantages with these large flocks. Hutto (1988) felt, because the foraging locations and rates of progression while feeding differed among species, some species must be making continual adjustments to match the overall rate of flock progression at the cost of feeding efficiency. Some of the core species are loud and give alarm calls and some are loud and also steal prey captured by other flockmates. These activities may be coordinated. The loud and aggressive White-winged Shrike-tanager, *Lanio versicolor*, a core species in canopy flocks, produces alarm calls to frighten other birds into dropping their newly captured prey item (Munn 1986). There is much more to be learned about the social interactions, costs, and benefits of mixed species flocks in the tropics.

The composition and interactions of flock participants is complex but flocks are fragile. Flocks disappear when forest structure is changed or fragmented. The first birds to disappear from fragmented tropical forests are obligate army ant followers (Willis and Oniki 1978, Lovejoy *et al.* 1986). When the forest area is too small to support the army ants, the birds that depend upon them disappear as well. Moving up from ant followers at ground level, the mid-level flocks of insectivorous and the canopy insectivorous and frugivorous flocks degrade or disappear with habitat alteration (Rappole and Morton 1985). So do those birds dependent upon the flocks. The destruction or even alteration of a tropical forest reduces avian biodiversity. But is biodiversity our most upsetting concern? To us, the answer is no.

7.5   Biodiversity or biotic interactions? What biotic interaction means to the conservation of tropical birds

Conservation biologists seem obsessed with the loss of taxa at phylogenetic levels above that of the species. Biological diversity, or biodiversity, is being increasingly viewed as a cladistic phenomenon (e.g. Nee and May 1997). Biodiversity is measured as the number of higher taxa (genera) and the total phylogenetic branch length, which is termed phylogenetic diversity (Purvis *et al.* 2000). Evolutionary history is equated with the total length of all the branches of the tree of life. We feel that something important is missing from this aspect of the loss of evolutionary history. At one level, what is missing is a feeling for the adaptations of animals at the species or population level. At another level, this view implies that what is important is cataloging the existence of species before they disappear. This has important implications for how effort is expended in this time of human rampage over the planet.

Latitudinal differences in biotic interactions suggests that conservation is of paramount importance in tropical regions for a reason rarely considered by conservationists. Even cladists agree that Darwin's theory can be applied in the modern world where we can actually see ecological relationships at work (Gee 1999). The simplification of habitats by humans *will be more devastating in the tropics than in the temperate zone* because biotic interactions have shaped a behavioral and morphological diversity in tropical birds that is far richer than that found in temperate zone birds. Biotic interactions produce complex evolutionary results of the sort most interesting to behavioral ecologists. Understanding these results within the huge behavioral diversity of tropical birds requires that the selection pressures underlying the traits can be inferred from current processes. With the alarming loss and degradation of tropical habitats we lose not just the individuals of a given species, but also the ability to study and understand the remarkable adaptations represented through these species. The adaptations themselves have not been described let alone the underlying selection balances that produced them. The strong biotic selection pressures mean that disruption of the environment and loss of species can quickly erase the evidence necessary to piece together evolutionary processes in the tropics.

This component of evolutionary history is more fragile than the phylogenetic branch lengths of cladists and has not been appreciated in science funding. The research that tropical behavioral ecologists are producing now will be historically important far beyond its present value. Recently, behavioral ecologists have begun to realize their science lacks a foundation in how behavior works, its pragmatic effects. Will this rising interest in the mechanisms of behavior be thwarted by the disappearance and distortion of the ecological theater? Surely research on biotic interactions and behavioral ecology of tropical birds should be top priority for funding and for positions, but we hear that Nero did fiddle as Rome burned.

References

Alatalo, R. V., C. Glynn and A. Lundburg. 1990. Singing rate and female attraction in the pied flycatcher: an experiment. *Anim. Behav.* **39**: 601–603.

Almeida, J. B. and R. H. Macedo. 2001. Lek-like mating system of the monogamous blue-black grassquit. *Auk*, in press.

Amundsen, T. 2000. Why are female birds ornamented? *Trends Ecol. Evol.* **15**: 149–155.

Amundsen, T., E. Forsgren and L. T. T. Hansen. 1997. On the function of female ornaments: male bluethroats prefer colorful females. *Proc. Royal Soc. Lond. B* **264**: 1579–1586.

Andersson, M. 1994. *Sexual selection.* Princeton Univ. Press, Princeton, NJ.

Arcese, P. 1987. Age, intrusion pressure, and defence against floaters by territorial male song sparrows. *Anim. Behav.* **35**: 773–784.

Arcese, P. 1989. Territory acquisition and loss in male song sparrows. *Anim. Behav.* **37**: 45–55.

Ashmole, N. P. 1963. The regulation of numbers of tropical oceanic birds. *Ibis* **103**: 458–473.

Austad, S. N. and K. N. Rabenold. 1985. Reproductive enhancement by helpers and an experimental examination of its mechanism in the bicolored wren: a facultatively communal breeder. *Behav. Ecol. Sociobiol.* **17**: 19–27.

Austen, M. J. W. and P. T. Handford. 1991. Variation in the songs of breeding Gambel's white-crowned sparrows near Churchill, Manitoba. *Condor* **93**: 147–152.

Bailey, S. F. 1978. Latitudinal gradients in colors and patterns of passerine birds. *Condor* **80**: 372–381.

Baker, J. R. 1938. The relation between latitude and breeding seasons in birds. *Proc. Zool. Soc. A* **108**: 557–582.

Bancroft, G. T., R. Bowman and R. J. Sawicki. 2000. Rainfall, fruiting phenology, and the nesting season of White-crowned Pigeons in the upper Florida Keys. *Auk* **117**: 416–426.

Baptista, L. F. 1975. Song dialects and demes in sedentary populations of the white-crowned sparrow, (*Zonotrichia leucophrys nuttali*). *Univ. of California Publ. Zool.* **105**: 1–52.

Bates, H. W. 1863. *The naturalist on the river Amazon.* Murray, London.

Beecher, M. D., S. E. Campbell, J. M. Burt, C. E. Hill and J. C. Nordby. 2000a. Song-type matching between neighbouring song sparrows. *Anim.*

Behav. **59**: 21–27.

Beecher, M. D., S. E. Campbell and J. C. Nordby. 2000b. Territory tenure in song sparrows is related to song sharing with neighbours, but not to repertoire size. *Anim. Behav.* 59: 29–37.

Beehler, B. 1983. Frugivory and polygamy in birds of paradise. *Auk* **100**: 1–12.

Beehler, B. 1985. Adaptive significance of monogamy in the Trumpet Manucode *Manucodia keraudrenii* (Aves: Paradisaeiidae). *Ornithol. Monogr.* **37**: 83–99.

Beehler, B. and S. G. Pruett-Jones. 1983. Display dispersion and diet of birds of paradise: a comparison of nine species. *Behav. Ecol. Sociobiol.* **13**: 229–238.

Beissinger, S. R. 1990. Experimental brood manipulations and the monoparental threshold in Snail Kites. *Amer. Nat.* **136**: 20–38.

Beissinger, S. R. and J. R. Waltman. 1991. Extraordinary clutch size and hatching asynchrony of a neotropical parrot. *Auk* **108**: 863–871.

Beletsky, L. 1996. *The red-winged blackbird, the biology of a strongly polygynous songbird.* Academic Press, San Diego.

Beletsky, L. D. and G. H. Orians. 1987. Territoriality among male redwinged blackbirds II. Removal experiments and site dominance. *Behav. Ecol. Sociobiol.* **20**: 339–349.

Beletsky, L. D. and G. H. Orians. 1989a. Territoriality among male redwinged blackbirds III. Testing hypotheses of territorial dominance. *Behav. Ecol. Sociobiol.* **24**: 333–339.

Beletsky, L. D., G. H. Orions and J. C. Wingfield. 1989b. Relationships of steroid hormones and polygyny to territorial status, breeding experience, and reproductive success in male red-winged blackbirds. *Auk* **106**: 107–117.

Beletsky, L. D., D. F. Gori, S. Freeman and J. C. Wingfield. 1995. Testosterone and polygyny in birds. *Curr. Ornithol.* **12**: 1–41.

Bennett, A. T. D., I. C. Cuthill, J. C. Partridge and E. J. Maier. 1996. Ultraviolet vision and mate choice in zebra finches. *Nature* **380**: 433–435.

Birkhead, T. R. 1998. Sperm competition in birds: mechanisms and function. Pp 579–622 In: *Sperm competition and sexual selection* (T.R. Birkhead and A. P. Møller, Eds). Academic Press, London.

Birkhead, T. R. and J. D. Biggins. 1987. Reproductive synchrony and extrapair copulation in birds. *Ethology* **74**: 320–334.

Birkhead, T. R. and K. Clarkson. 1985. Ceremonial gatherings of the magpie *Pica pica*: territory probing and acquisition. *Behaviour* **94**: 324–332.

Birkhead, T. R. and A. P. Møller. 1992. *Sperm competition in birds: evolutionary causes and consequences.* Academic Press, London.

Birkhead, T. R. and A. P. Møller. 1996. Monogamy and sperm competition in birds. Pp 323–343 In: *Partnerships in birds* (J. M. Black, Ed.). Oxford Univ. Press, Oxford.

Blake, E. R. 1953. *Birds of Mexico.* Univ. of Chicago Press, Chicago.

Bleiweiss, R. 1985. Iridescent polychromatism in a female hummingbird: Is it related to feeding strategies? *Auk* **102**: 701–713.

Bleiweiss, R. 1997. Covariation of sexual dichromatism and plumage colours in lekking and non-lekking birds: a comparative analysis. *Evol. Ecol.* **11**: 217–235.

Boag, P. R. and P. R. Grant. 1984. Darwin's Finches (Geospiza) on Isla Daphne Major, Galapagos: breeding and feeding ecology in a climatically variable environment. *Ecol. Monogr.* **54**: 463–489.

Bradbury, J. W. 1981. The evolution of leks. Pp 138–169 In: *Natural selection and social behavior* (R. D. Alexander and D. Tinkle, Eds). Chiron Press, New York.

Breitwisch, R., P. G. Merritt and G. H. Whitesides. 1984. Why do Northern Mockingbirds feed fruit to their nestlings? *Condor* **86**: 281–287.

Brenowitz, E. A., K. Lent and D. E. Kroodsma. 1995. Brain space for learned song in birds develops independently of song learning. *J. Neurosci.* **15**: 6281–6286.

Brown, J. L. 1987. *Helping and communal breeding in birds.* Princeton Univ. Press, Princeton.

Brown, J. L., E. R. Brown, J. Sedransk and S. Ritter. 1997. Dominance, age, and reproductive success in a complex society: a long-term study of the Mexican jay. *Auk* **114**: 279–286.

Burley, N. 1988. The differential allocation hypothesis: an experimental test. *Amer. Nat.* **132**: 611–628.

Burns, K. 1998. A phylogenetic perspective on the evolution of sexual dichromatism in tanagers (Thraupidae): the role of female versus male plumage. *Evolution* **52**: 1219–1224.

Buskirk, W. H. 1976. Social systems in a tropical forest avifauna. *Amer. Nat.* **110**: 293–310.

Catchpole, C. K. and B. Leisler. 1996. Female aquatic warblers (*Acrocephalus paludicola*) are attracted by playback of longer and more complicated songs. *Behaviour* **133**: 1153–1164.

Charnov, E. R. and J. R. Krebs. 1975. The evolution of alarm calls: altruism or manipulation? *Am. Nat.* **109**: 107–112.

Chen, D. M. and T. H. Goldsmith. 1986. Four spectral classes of cone in the retinas of birds. *J. Comp. Phys.* A **159**: 473–479.

Cipollini, M. L. and E. W. Stiles. 1993a. Fruit rot, antifungal defense, and palatability of fleshy fruits for frugivorous birds. *Ecology* **74**: 751–762.

Cipollini, M. L. and E. W. Stiles. 1993b. Fungi as biotic defense agents of fleshy fruits: alternative hypotheses, predictions, and evidence. *Am. Nat.* **141**: 663–673.

Cockburn, A. 1998. Evolution of helping behavior in cooperatively breeding birds. *Ann. Rev. Ecol. Syst.* **29**: 141–177.

Cody, M. L. 1966. A general theory of clutch size. *Evolution* **20**: 174–184.

Conrad, K. E., M. F. Clarke, R. J. Robertson and P. T. Boag. 1998. Paternity and the relatedness of helpers in the cooperatively breeding bell miner. *Condor* **100**: 343–349.

Cott, H. B. 1940. *Adaptive coloration in animals.* Methuen, London.

Crook, J. H. 1964. The evolution of social organization and visual communication in weaverbirds (Ploceinae). *Behaviour* (Suppl) **10**: 1–178.

Cunningham, E. J. A. and T. R. Birkhead. 1998. Sex roles and sexual selection. *Anim. Behav.* **56**: 1311–1321.

Curry, R. L. and P. R. Grant. 1990. Galapagos mockingbirds: territorial cooperative breeding in a climatically variable environment. Pp 291–329 In: *Cooperative breeding in birds* (P. B. Stacey and W. D. Koenig, Eds). Cambridge Univ. Press, Cambridge.

Cuthill, I. C. and W. A. Macdonald. 1990. Experimental manipulation of the dawn and dusk chorus in the blackbird (*Turdus merula*). *Behav. Ecol. Sociobiol.* **26**: 209–216.

Cuthill, I. C., A. T. D. Bennett, J. C. Partridge and E. J. Maier. 1999. Plumage reflectance and the objective assessment of avian sexual dichromatism. *Amer. Nat.* **160**: 183–200.

Davidar, P. 1983. Birds and neotropical mistletoes: effects of seedling recruitment. *Oecologia* **60**: 271–273.

Davidar, P. 1987. Fruit structure in mistletoes and its consequences for seed dispersal. *Biotropica* **19**: 137–139.

Davidar, P. and E. S. Morton. 1986. On the relationship between fruit crop size and fruit removal by birds. *Ecology* **76**: 262–265.

Davidse, G. and E. S. Morton. 1973. Bird-mediated fruit dispersal in the tropical genus *Lasiacis* (Graminae: panicaae). *Biotropica* **5**: 162–167.

Dawson, W. R. and F. C. Evans. 1957. Relation of growth and development to temperature regulation in nestling field and chipping sparrows. *Physiol. Zoöl.* **30**: 315–327.

Deerenberg, C., V. Arpanius, S. Daan and N. Bos. 1997. Reproductive effort decreases antibody responsiveness. *Proc. Royal. Soc. Lond.* B **264**: 1021–1029.

deSchauensee, R. M. 1964. *Birds of Columbia*. Livingston Publ. Co. Narberth PA.

deSchauensee, R. M. and W. H. Phelps, Jr. 1978. *A guide to the Birds of Venezuela*. Princeton Univ. Press.

Dittami, J. P. and E. Gwinner. 1990. Endocrine correlates of seasonal reproduction and territorial behavior in some tropical passerines. Pp 225–233 In: *Endocrinology of birds: molecular to behavioral* (M. Wada, Ed.). Japan Science Society and Springer, Tokyo and Berlin.

Dooling, R. J. 1982. Auditory perception in birds. Pp 95–130 In: *Acoustic communication in birds*, vol. 1 (D. E. Kroodsma and E. H. Miller, Eds). Academic Press, New York.

Dumbacker, J. P., B. M. Beehler, T. F. Spande, H. M. Garraffo and J. W. Daly. 1992. Homobatrachotoxin in the genus *Pitohui*: chemical defense in birds? *Science* **258**: 799–801.

Dyrcz, A. 1983. Breeding ecology of the clay-coloured robin *Turdus grayi* in lowland Panama. *Ibis* **125**: 287–304.

Eckman, J. 1988. Subordination costs and group territoriality in wintering willow tits. *Proc. Int. Ornithol. Congr.* **14**: 2372–2381.

Elekonich, M. M. 2000. Female song sparrow, *Melospiza melodia*, response to simulated conspecific and heterospecific intrusion across three seasons. *Anim. Behav.* **29**: 551–557.

Emlen, S. T. 1981. Altruism, kinship, and reciprocity in the White-fronted Bee-eater. Pp 217–320 In: *Natural selection and social behavior* (R. D. Alexander and D. W. Tinkle, Eds). Chiron Press, New York.

Emlen, S. T. 1982. The evolution of helping. I. An ecological constraints model. *Amer. Nat.* **119**: 29–39.

Emlen, S. T. and L. W. Oring. 1977. Ecology, sexual selection, and the evolution of mating systems. *Science* **197**: 215–223.

Emlen, S. T., P. H. Wrege and M. S. Webster. 1998. Cuckoldry as a cost of polyandry in the sex-role-reversed wattled jacana, *Jacana jacana*. *Proc. Royal Soc. Lond* **265**: 2359–2364.

Endler, J. A. 1978. A predator's view of animal color patterns. *Evol. Biol.* **11**: 319–364.

Endler, J. A. and M. Théry. 1996. Interacting effects of lek placement, display behavior, ambient light, and color patterns in three Neotropical forest-dwelling birds. *Amer. Nat.* **148**: 421–452.

Enstrom, D. A. 1993. Female choice for age-specific plumage in the orchard oriole: implications for delayed plumage maturation. *Anim. Behav.* **45**: 435–442.

Estrada, A. and T. Fleming. 1985. *Frugivores and seed dispersal*. Junk, The Hague.

Etchecopar, R. D. and F. Hue. 1967. *The Birds of North Africa*. Oliver & Boyd, London.

Evans Ogden, L. G. and B. J. M. Stutchbury. 1996. Constraints on double brooding in a neotropical migrant, the Hooded Warbler. *Condor* **98**: 736–744.

Evans Ogden, L .G. and B. J. M. Stutchbury. 1997. Fledgling care and male parental effort in the Hooded Warbler (*Wilsonia citrina*). *Can. J. Zool.* **75**: 576–581.

Faaborg, J. 1986. Reproductive success and survivorship in the Galapagos Hawk, *Buteo galapagoensis*: potential costs and benefits of cooperative polyandry. *Ibis* **128**: 337–347.

Faaborg, J. and W. J. Arendt. 1995. Survival rates of Puerto Rican birds: are islands really that different? *Auk* **112**: 503–507.

Falls, J. B., J. R. Krebs and P. K. Mcgregor. 1982. Song matching in the Great tit (*Parus major*): the effect of similarity and familiarity. *Anim. Behav.* **30**: 977–1009.

Farabaugh, S. M. 1982. The ecological and social significance of duetting. Pp 85–124 In: *Acoustic communication in birds* (D. E. Kroodsma and E. H. Miller, Eds). Academic Press, New York.

Feekes, F. 1981. Biology and colonial organization of two sympatric caciques, *Cacicus c. cela* and *Cacicus h. haemorrhous* (Icteridae, Aves) in Surinam. *Ardea* **69**: 83–107.

Fleischer, R. C., C. Tarr and T. K. Pratt. 1994. Genetic structure and mating system in the palila, an endangered Hawaiian honeycreeper, as assessed by DNA fingerprinting. *Molecular Ecology* **3**: 383–392.

Fleischer, R. C., C. Tarr, E. S. Morton, K. D. Derrickson and A. Sangmeister. 1997. Mating system of the Dusky Antbird (*Cercomacra tyrannina*), a tropical passerine, as assessed by DNA fingerprinting. *Condor* **99**: 512–514.

Fleming, T. H., R. Breitswich and G. H. Whitesides. 1987. Patterns of tropical vertebrate diversity. *Ann. Rev. Ecol. Syst.* **18**: 91–109.

Fogden, M. P. L. 1972. The seasonality and population dynamics of equatorial forest birds in Sarawak. *Ibis* **114**: 307–343.

Folstad, I. and A. J. Karter. 1992. Parasites, bright males, and the immunocompetence handicap. *Amer. Nat.* **139**: 603–622.

Forsyth, A. and K. Miyata. 1984. *Tropical nature. Life and death in the rainforests of Central and South America.* Simon and Schuster Publ., New York.

Foster, M. S. 1974. A model to explain molt-breeding overlap and clutch size in some tropical birds. *Evolution* **28**: 182–190.

Foster, M. S. 1975. The overlap of molting and breeding in some tropical birds. *Condor* **77**: 304–314.

Foster, M. S. 1981. Cooperative behavior and social organization of the Swallow-tailed Manakin (*Chiroxiphia caudata*). *Behav. Ecol. Sociobiol.* **9**: 167–177.

Foster, M. S. 1987a. Feeding methods and efficiencies of selected frugivorous birds. *Condor* **89**: 566–580.

Foster, M. S. 1987b. Delayed maturation, neoteny, and social system differences in two manakins of the genus *Chiroxiphia*. *Evolution* **41**: 547–558.

Foster, M. and R. W. McDiarmid. 1983. Nutritional value of the aril of *Trichilia cuneata*, a bird-dispersed fruit. *Biotropica* **15**: 26–31.

Foster, R. B. 1982. The seasonal rhythm of fruitfall on Barro Colorado Island. In: *The ecology of a tropical rainforest* (E. G. Leigh, Jr., A. S. Rand and D. M. Windsor, Eds). Smithsonian Inst. Press, Washington.

Fotheringham, J. R., P. R. Martin and L. Ratcliffe. 1997. Song transmission and auditory perception of distance in wood warblers (Parulinae). *Anim. Behav.* **53**: 1271–1285.

Freed, L. A. 1986. Territory takeover and sexually selected infanticide in tropical house wrens. *Behav. Ecol. Sociobiol.* **19**:197–206.

Freed, L. A. 1987. The long-term pair bond of tropical house wrens: advantage or constraint? *Amer. Nat.* **130**: 507–525.

Fuentes, M. 1992. Latitudinal and elevational variation in fruiting phenology among western European bird-dispersed plants. *Ecography* **15**: 177–183.

Gee, H. 1999. *In search of deep time: beyond the fossil record to a new history of life.* The Free Press, New York.

Geffen, E. and Y. Yom-Tov. 2000. Are incubation and fledging periods longer in the tropics? *J. Anim. Ecol.* **69**: 59–73.

Gibbs, J. P. 1991. Avian nest predation in tropical wet forest: an experimental study. *Oikos* **60**: 155–161.

Gil, D., J. A. Graves and P. J. B. Slater. 1999. Seasonal patterns of singing in the willow warbler: evidence against the fertility announcement hypothesis. *Anim. Behav.* **58**: 995–1000.

Gilbert, W. M. and A. F. Carroll. 1999. Singing in a mated female Wilson's Warbler. *Wilson Bull.* **111**: 134–137.

Gottlander, K. 1987. Variation in the song rate of the male pied flycatcher (*Ficedula hypoleuca*): causes and consequences. *Anim. Behav.* **35**: 1037–1043.

Gradwhol, J. and R. Greenberg. 1980. The formation of antwren flocks on Barro Colorado Island, Panama. *Auk* **97**: 385–395.

Gradwohl, J. and R. Greenberg. 1982. The breeding season of antwrens on Barro Colorado Island. Pp 345–351. In: *The ecology of a tropical rainforest* (E. G. Leigh, Jr., A. S. Rand and D. M. Windsor, Eds). Smithsonian Inst. Press, Washington.

Grant, P. R. and B. R. Grant. 1992. Demography and the genetically effective sizes of two populations of Darwin's Finches. *Ecology* **73**: 766–784.

Graves, G. R. and N. J. Gotelli. 1993. Assembly of avian mixed-species flocks in Amazonia. *Proc. Natl. Acad. Sci. USA* **90**: 1388–1391.

Greenberg, J. 1983. Natural highs in natural habitats. *Science News* **124**: 300–301.

Greenberg, R. 1981. Frugivory in some migrant tropical forest wood warblers. *Biotropica* **13**: 215–223.

Greenberg, R. 1983. The role of neophobia in determining the degree of foraging specialization in some migrant warblers. *Am. Nat.* **122**: 444–453.

Greenberg, R. 1984a. The winter exploitation system of Bay-breasted and Chestnut-sided Warblers in Panama. *Univ. Calif. Publ. Zool.* 116. University of California Press, Berkeley.

Greenberg, R. 1984b. The role of neophobia in the foraging selection of a tropical migrant bird: an experimental study. *Proc. Natl. Acad. Sci USA* **81**: 3778–3780.

Greenberg, R. 1987. Development of dead leaf foraging in a tropical migrant warbler. *Ecology* **68**: 130–141.

Greenberg, R. and J. Gradwohl. 1983. Sexual roles in the Dot-winged Antwren (*Microrhopias quixensis*), a tropical forest passerine. *Auk* **100**: 920–925.

Greenberg, R. and J. Gradwohl. 1986. Constant density and stable territoriality in some tropical insectivorous birds. *Oecologia* **69**: 618–625.

Greenberg, R. and J. Gradwohl. 1997. Territoriality, adult survival, and dispersal in the Checker-throated Antwren in Panama. *J. Avian. Biol.* **28**: 103–110.

Greenberg, R., D. K. Niven, S. Hopp and C. Boone. 1993. Frugivory and coexistence in a resident and a migratory vireo on the Yucatan Peninsula. *Condor* **95**: 990–999.

Greenberg, R., M. Foster and L. Marquez. 1995. The role of White-eyed Vireos in the dispersal of *Bursera simaruba* fruit. *J. Trop. Ecol.* **11**: 619–639.

Griscom, L. and A. Sprunt. 1957. *The warblers of America*. Devin-Adair Co., New York.

Haggerty, T., E. S. Morton and R. C. Fleischer. 2001. Genetic monogamy in the Carolina Wren, *Thryothorus ludovicianus. Auk*, in press.

Hails, C. J. 1982. A comparison of tropical and temperate aerial insect abundance. *Biotropica* **14**: 310–313.

Halkin, S. L. 1997. Nest-vicinity song exchanges may coordinate biparental care of northern cardinals. *Anim. Behav.* **54**: 189–198.

Hamilton, W. D. and M. Zuk. 1982. Heritable true fitness and bright birds: a role for parasites? *Science* **218**: 384–387.

Hansen, A. J. and S. Rohwer. 1986. Coverable badges and resource defence in birds. *Anim. Behav.* **34**: 69–76.

Harris, M. P. 1970. Breeding ecology of the Swallow-tailed Gull, *Creagrus furcatus. Auk* **87**: 215–243.

Harris, M. A. and R. E. Lemon. 1974. Songs of Song Sparrows (*Melospiza melodia*): individual variation and dialects. *Can. J. Zool.* **50**: 301–309.

Hasselquist, D., S. Bensch and T. von Schantz. 1996. Correlation between song repertoire, extra-pair paternity and offspring survival in the great reed warbler. *Nature* **381**: 229–232.

Hasselquist, D., J. A. Marsh, P. W. Sherman, and J. C. Wingfield. 1999. Is avian humoral immunocompetence suppressed by testosterone? *Behav. Ecol. Sociobiol.* **45**: 167–175.

Hau, M., M. Wikelski and J. C. Wingfield. 1998. A neotropical forest bird can measure the slight changes in tropical photoperiod. *Proc. R. Soc. Lond.* B **265**: 89–95.

Hau, M., M. Wikelski, K. K. Soma and J. C. Wingfield. 2000. Testosterone and year-round territorial aggression in a tropical bird. *Gen. Comp. Endocrinol.* **117**: 20–33.

Haverschmidt, F. 1968. *Birds of Surinam*. Oliver and Boyd, London.

Haydock, J., P. G. Parker and K. N. Rabenold. 1996. Extra-pair paternity is uncommon in the cooperatively breeding Bicolored Wren. *Behav. Ecol. Sociobiol.* **38**: 1–16.

Herrera, C. M. 1981. Are tropical fruits more rewarding to their dispersers than temperate ones? *Am. Nat.* **118**: 896–907.

Hill, G. E. 1990. Female house finches prefer colourful males: sexual selection for a condition-dependent trait. *Anim. Behav.* **40**: 563–572.

Hill, G. E. 1993. Male mate choice and the evolution of female plumage coloration in the House Finch. *Evolution* **47**: 1515–1525.

Hoi, H. 1997. Assessment of the quality of copulation partners in the monogamous bearded tit. *Anim. Behav.* **53**: 277–286.

Hoi, H. and M. Hoi-Leitner. 1997. An alternative route to coloniality in the bearded tit: females pursue extra-pair fertilizations. *Behav. Ecol.* **8**: 113–119.

Hoi-Leitner, M., H. Nechtelberger and H. Hoi. 1995. Song rate as a signal for nest site quality in blackcaps (*Sylvia atricapilla*). *Behav. Ecol. Sociobiol.* **37**: 399–405.

Holmes, R. T. and J. C. Schultz. 1988. Food availability for forest birds: effects of prey distribution and abundance on bird foraging. *Can. J. Zool.*

66: 720–728.

Holmes, R. T., T. W. Sherry and F. W. Sturges. 1986. Bird community dynamics in a temperate deciduous forest: long-term trends at Hubbard Brook. *Ecol. Monogr.* **56**: 201–220.

Howe, H. F. 1979. Fear and frugivory. *Am. Nat.* **114**: 925–931.

Howe, H. F. and G. F. Estabrook. 1977. On intraspecific competition for avian dispersers in tropical trees. *Am. Nat.* **111**: 817–832.

Howe, H. F. and J. Smallwood. 1982. Ecology of seed dispersal. *Ann. Rev. Ecol. Syst.* **13**: 201–228.

Hunt, S., I. C. Cuthill, A. T. D. Bennett and R. Griffiths. 1999. Preferences for ultraviolet partners in the blue tit. *Anim. Behav.* **58**: 809–815.

Hutchinson, G. E. 1965. *The ecological theater and the evolutionary play.* Yale University Press, New Haven.

Hutto, R. L. 1988. Foraging behavior patterns suggest a possible cost associated with participation in mixed-species flocks. *Oikos* **51**: 79–83.

Innes, K. E. and R. E. Johnston. 1996. Cooperative breeding in the white-throated magpie-jay. How do auxiliaries influence nesting success? *Anim. Behav.* **51**: 519–533.

Irwin, R. E. 1994. The evolution of plumage dichromatism in the New World blackbirds: social selection on female brightness? *Amer. Nat.* **144**: 890–907.

Isler, M. L. and P. R. Isler. 1999. *The Tanagers.* Smithsonian Inst. Press., Washington.

Janzen, D. H. 1969. Birds and the ant x acacia interaction in Central America, with notes on birds and other myrmecophytes. *Condor* **71**: 240–256.

Janzen, D. H. 1973. Sweep samples of tropical foliage insects: description of study sites with data on species abundances and size distributions. *Ecology* **54**: 659–686.

Janzen, D. H. 1975. *Ecology of plants in the tropics.* Edward Arnold, London.

Janzen, D. H. 1980. When is it coevolution? *Evolution* **34**: 611–612.

Johnsen, A., S. Andersson, J. Ornborg, and J. T. Lifjeld. 1998a. Ultraviolet plumage ornamentation affects social mate choice and sperm competition in bluethroats (Aves: *Luscinia s. svecica*): a field experiment. *Proc. Royal. Soc. Lond.* B **265**: 1313–1318.

Johnsen, A., J. T. Lifjeld, P. A. Rohde, C. R. Primmer and H. Ellegren. 1998b. Sexual conflict over fertilizations: female bluethroats escape male paternity guards. *Behav. Ecol. Sociobiol.* **43**: 401–408.

Johnsen, T. S. 1998. Behavioral correlates of testosterone and seasonal changes in steroids in red-winged blackbirds. *Anim. Behav.* **55**: 957–965.

Johnsgard, P. A. 1994. *Arena Birds.* Smithsonian Inst. Press, Washington.

Johnson, K. P., F. McKinney and M. D. Sorenson. 1998. Phylogenetic constraint on male parental care in the dabbling ducks. *Proc. Royal Soc. Lond* B **266**: 759–763.

Johnston, J. P., W. J. Peach, R. D. Gregory and S. A. White. 1997. Survival rates of temperate and tropical passerines: A Trinidadian perspective. *Am.*

Nat. **150**: 771–789.

Jones, I. L. and F. M. Hunter. 1993. Mutual sexual selection in a monogamous seabird. *Nature* **362**: 238–239.

Jones, I. L. and F. M. Hunter. 1999. Experimental evidence for mutual inter- and intrasexual selection favouring a crested auklet ornamentation. *Anim. Behav.* **57**: 521–528.

Joyce, F. J. 1993. Nesting success of rufous-naped wrens (*Campylorhynchus rufinucha*) is greater near wasp nests. *Behav. Ecol. Sociobiol.* **32**: 71–77.

Karr, J. R., J. D. Nichols, M. K. Klimkiewicz and J. D. Brawn. 1990. Survival rates of birds of tropical and temperate forests: will the dogma survive? *Amer. Nat.* **136**: 277–291.

Kempenaers, B. 1993. The uses of a breeding synchrony index. *Ornis. Scand.* **24**: 84.

Kempenaers, B., G. R. Verheyen and A. A. Dhont. 1995. Mate guarding and copulation behavior in monogamous and polygynous Blue Tits: do males follow a best-of-a-bad job? *Behav. Ecol. Sociobiol.* **36**: 33–42.

Kempenaers, B., G. R. Verheyen, M. Van den Broeck, T. Burke, C. Van Broeckhoven and A. A. Dhondt. 1992. Extra-pair paternity results from female preference for high-quality males in the blue tit. *Nature* **357**: 494–496.

Kennedy, R. S., P. C. Gonzales and H. C. Miranda, Jr. 1997. New *Aethopyga* Sunbirds (Aves: Nectariniidae) from the island of Mindanao, Phillipines. *Auk* **114**: 1–10.

Ketterson, E. D. and V. Nolan, Jr. 1994. Parental behavior in birds. *Ann. Rev. Ecol. Syst.* **25**: 601–628.

Ketterson, E. D., V. Nolan, Jr., L. Wolf and C. Ziegenfus. 1992. Testosterone and avian life histories: effects of experimentally elevated testosterone on behavior and correlates of fitness in the dark-eyed junco (*Junco hyemalis*). *Amer. Nat.* **140**: 980–999.

Klomp, H. 1970. The determination of clutch-size in birds. A review. *Ardea* **58**: 1–124.

Koenig, W. D. and F. A. Pitelka. 1981. Ecological factors and kin selection in the evolution of cooperative breeding in birds. Pp 261–280 In: *Natural selection and social behavior* (R. D. Alexander and D. W. Tinkle, Eds). Chiron Press, New York.

Komdeur, J. A. 1992. Importance of habitat saturation and territory quality for evolution of cooperative breeding in the Seychelles Warbler. *Nature* **358**: 493–495.

Komdeur, J. A. 1996. Seasonal timing of reproduction in a tropical bird, the Seychelles warbler: a field experiment using translocation. *J. Biol. Rhythms* **11**: 333–350.

Komdeur, J. A., A. Huffstadt, W. Prast, G. Castle, R. Mileto and J. Wattel. 1995. Transfer experiments of Seychelles Warblers to new islands: changes in dispersal and helping behaviour. *Anim. Behav.* **49**: 695–708.

Konishi, M. 1969. Time resolution by single auditory neurones in birds. *Nature* **222**: 566–567.

Krebs, J. R. 1982. Territorial defense in the great tit (*Parus major*): do residents always win? *Behav. Ecol. Sociobiol.* **11**: 185–194.

Krebs, J. R. and N. B. Davies. 1991. *Behavioral Ecology*. 3rd Ed. Blackwell Scientific Publications, Oxford.

Krebs, J. R. and D. E. Kroodsma. 1980. Repertoires and geographical variation in bird song. *Adv. Stud. Behav.* **11**: 143–177.

Krebs, J. R., R. Ashcroft and K. V. Orsdol. 1981. Song matching in the great tit, *Parus major. Anim. Behav.* **29**: 919–923.

Kricher, J. 1997. *A neotropical companion*. 2nd ed. Princeton Univ. Press, Princeton.

Krokene, C., K. Anthonisen, J. T. Lifjeld and T. Amundsen. 1996. Paternity and paternity assurance behaviour in the bluethroat, *Luscinia s. svecica. Anim. Behav.* **52**: 405–417.

Kroodsma, D. E., W-C Liu, E. Goodwin and P. A. Bedell. 1999. The ecology of song improvisation as illustrated by North American sedge wrens. *Auk* **116**: 373–386.

Kulesza, G. 1990. An analysis of clutch-size in New World passerine birds. *Ibis* **132**: 407–422.

Kunkel, P. 1974. Mating systems of tropical birds: the effects of weakness or absence of external reproduction-timing factors, with special reference to prolonged pair bonds. *Z. Tierpsychol.* **34**: 265–307.

Lack, D. 1947. The significance of clutch-size. *Ibis* **89**: 302–352.

Lack, D. 1948. The significance of clutch-size. *Ibis* **90**: 24–45.

Lack, D. 1950. The breeding seasons of European birds. *Ibis* **92**: 288–316.

Lack, D. 1954. *The natural regulation of animal numbers*. Clarendon Press, Oxford.

Lack, D. 1968. *Ecological adaptations for breeding in birds*. Metheun, London.

Lack, D. and R. E. Moreau. 1965. Clutch-size in tropical passerine birds of forest and savannah. *Oiseau* **35**: 76–89.

Lambrechts, M. M., P. Perret and J. Blondel. 1996. Adaptive differences in the timing of egg laying between different populations of birds result from variation in photoresponsiveness. *Proc. R. Soc. Lond* B **163**: 19–22.

Langen, T. A. and S. L. Vehrencamp. 1998. Ecological factors affecting group size and territory size in white-throated magpie-jays. *Auk* **115**: 327–339.

Lefebvre, G., B. Poulin and R. McNeil. 1992. Settlement period and function of long-term territory in tropical mangrove passerines. *Condor* **94**: 83–92.

Lepson, J. K. and L. A. Freed. 1995. Variation in male plumage and behavior of the Hawaii Akepa. *Auk* **112**: 402–414.

Lessells, C. M. 1991. The evolution of life histories. Pp 32–65 In: *Behavioural ecology, an evolutionary approach* (J. R. Krebs and N. B. Davies, Eds). 3rd Ed. Blackwell, Oxford.

Levey, D. J. 1988. Spatial and temporal variation in Costa Rican fruit and fruit-eating bird abundance. *Ecol. Monogr.* **58**: 251–269.

Levey, D. J. and W. H. Karasov. 1989. Digestive responses of temperate birds switched to fruit or insect diets. *Auk* **106**: 675–686.

Levey, D. J. and W. H. Karasov. 1992. Digestive modulation in a seasonal frugivore, the American robin. *Am. J. Physiol.* **262**: G711–G718.

Levey, D. J. and F. G. Stiles. 1992. Evolutionary precursors of long-distance migration: resource availability and movement patterns in neotropical landbirds. *Am. Nat.* **140**: 447–476.

Levey, D. J. and F. G. Stiles. 1994. Birds: ecology, behavior and taxonomic affinities. Pp. 217–228 In: *La Selva. Ecology and natural history of a tropical rain forest* (L. A. Dade, K. S. Bawa, H. A. Hespenheide, and G. S. Hartshorn, Eds). Chicago Univ. Press, Chicago.

Levin, R. N. 1996a. Song behaviour and reproductive strategies in a duetting wren, *Thryothorus nigricapillus*. I. Removal experiments. *Anim. Behav.* **52**: 1093–1106.

Levin, R. N. 1996b. Song behaviour and reproductive strategies in a duetting wren, *Thryothorus nigricapillus*. II. Playback experiments. *Anim. Behav.* **52**: 1107–1117.

Levin, R. N. and J. C. Wingfield. 1992. The hormonal control of territorial aggression in tropical birds. *Ornis Scand.* **23**: 284–291.

Ligon, J. D. 1981. Demographic patterns and communal breeding in the Green Woodhoopoe *Phoeniculus purpureus*. Pp 231–243 In: *Natural selection and social behavior* (R. D. Alexander and D. W. Tinkle, Eds). Chiron Press, New York.

Ligon, J. D. and S. H. Ligon. 1990. Green woodhoopoes: life history traits and sociality. Pp 33–65 In: *Cooperative breeding in birds* (P. B. Stacey and W. D. Koenig, Eds). Cambridge Univ. Press, Cambridge.

Lill, A. 1974. The evolution of clutch size and male 'chauvinism' in the white-bearded manakin. *Living Bird* **13**: 211–231.

Lima, S. L. 1987. Clutch size in birds: a predation perspective. *Ecology* **68**: 1062–1070.

Lima, S. L. 1998. Stress and decision making under the risk of predation: recent developments from behavioral, reproductive, and ecological perspectives. *Adv. Study Behav.* **27**: 215–290.

Loiselle, B. A. and W. G. Hoppes. 1983. Nest predation in insular and mainland lowland rainforest in Panama. *Condor* **85**: 93–95.

Lovejoy, T. E., R. O. Bierregaard Jr., A. B. Rylands, J. R. Malcolm, C. E. Quintela, L. H. Harper, K. S. Brown Jr., A. H. Powell, G. V. N. Powell, H. O. R. Schubart and M. B. Hayes. 1986. Edge and other effects of isolation of Amazon forest fragments. Pp 257–285 In: *Conservation biology: the science of scarcity and diversity.* (M. E. Soule, Ed.). Sinauer, Sunderland, Mass.

Lyon, B. E. and R. D. Montgomerie. 1986. Delayed plumage maturation in passerine birds: reliable signaling by subordinate males? *Evolution* **40**: 604–615.

Mabey, S. and E. S. Morton. 1992. Demography and territorial behavior of wintering Kentucky warblers in Panama. Pp 329–336 In *Ecology and conservation of neotropical migrant landbirds* (J. M. Hagen and D. W. Johnston, Eds). Smithsonian Institution Press, Washington.

Mace, R. 1987. The dawn chorus in the great tit *Parus major* is directly related to female fertility. *Nature* **330**: 745–746.

Macedo, R. H. and C. A. Bianchi. 1997. Communal breeding in tropical Guira Cuckoos *Guira guira*: sociality in the absence of a saturated habitat. *J. Avian Biol.* **28**: 207–215.

MacDougall-Shackleton, E. A. and R. J. Robertson. 1998. Confidence of paternity and paternal care by eastern bluebirds. *Behav. Ecol.* **9**: 201–205.

Mader, W. J. 1982. Ecology and breeding habits of the Savannah Hawk in the llanos of Venezuela. *Condor* **84**: 261–271.

Magrath, M. J. L. and M. A. Elgar. 1997. Paternal care declines with increased opportunity for extra-pair matings in fairy martins. *Proc. Royal Soc. London* B **264**: 1731–1736.

Marchant, S. 1960. The breeding of some S. W. Ecuadorian birds. *Ibis* **102**: 349–382; 584–599.

Marler, P. 1999. On innateness: are sparrow songs 'learned' or 'innate'? Pp 293–318 In: *The Design of Animal Communication* (M. Hauser and M. Konishi, Eds). MIT Press, Cambridge, Massachusetts.

Marler, P. and M. Tamura. 1962. Song 'dialects' in three populations of white-crowned sparrows. *Condor* **64**: 368–377.

Martin, T. E. 1987. Food as a limit on breeding birds: a life-history perspective. *Ann. Rev. Ecol. Syst.* **18**: 453–487.

Martin, T. E. 1993. Nest predation among vegetation layers and habitat types: revising the dogmas. *Amer. Nat.* **141**: 897–913.

Martin, T. E. 1995. Avian life history evolution in relation to nest sites, nest predation, and food. *Ecol. Monogr.* **65**: 101–127.

Martin, T. E. 1996. Life history evolution in tropical and south temperate birds: what do we really know? *J. Avian Biol.* **27**: 263–272.

Martin, T. E. and A. Badyaev. 1996. Sexual dichromatism in birds: importance of nest predation and nest location for females versus males. *Evolution* **50**: 2454–2460.

Martin, T. E., P. R. Martin, C. R. Olson, B. J. Heidinger and J. J. Fontaine. 2000. Parental care and clutch sizes in North and South American birds. *Science* **287**: 1482–1485.

Martinez del Rio, C. and W. H. Karasov. 1990. Digestive strategies in nectar- and fruit-eating birds and the sugar composition of plant rewards. *Am. Nat.* **136**: 618–656.

Matthysen, E. 1989. Territorial and nonterritorial settling in juvenile Eurasian nuthatches (*Sitta europea* L.) in summer. *Auk* **106**: 560–567.

Mauck, R. A., T. A. Waite and P. G. Parker. 1995. Monogamy in Leach's Storm-petrel: DNA fingerprinting evidence. *Auk* **112**: 473–482.

Maynard Smith, J. and G. A. Parker. 1976. The logic of asymmetric contests. *Anim. Behav.* **24**: 159–175.

McDonald, D. B. 1989. Cooperation under sexual selection: age-graded changes in a lekking bird. *Amer. Nat.* **134**: 709–730.

McDonald, D. B. 1993. Demographic consequences of sexual selection in the long-tailed manakin. *Behav. Ecol.* **4**: 297–309.

McKey, D. 1975. The ecology of coevolved seed dispersal systems. Pp 159–191 In: *Coevolution of animals and plants* (L. E. Gilbert and P. R. Raven, Eds). University of Texas Press, Austin.

McKinney, F. 1985. Primary and secondary male reproductive strategies of dabbling ducks. Pp 68–82 In: *Avian Monogamy* (P. A. Gowaty and D. W. Mock, Eds). *Ornithol. Monogr.* 37, American Ornithologists Union, Washington DC.

McKinney, F. and G. Brewer. 1989. Parental attendance and brood care in four Argentine dabbling ducks. *Condor* 91: 131–138.

McKinney, F., S. R. Derrickson and P. Mineau. 1983. Forced copulation in waterfowl. *Behaviour* 86: 250–294.

Medsger, O. P. 1931. *Nature rambles. Spring.* Corwall Press, New York.

Melland, R. R. 2000. The genetic mating system and population structure of the Green-rumped Parrotlet. PhD Dissertation, Univ. of North Dakota, Grand Forks, ND.

Midgley, J. J. and W. J. Bond. 1991. Ecological aspects of the rise of angiosperms: a challenge to the reproductive superiority hypothesis. *Biol. J. Linn. Soc.* 44: 81–92.

Miller, A. H. 1962. Bimodal occurrence of breeding in an equatorial sparrow. *Proc. Natl. Acad. Sci.* 48: 396–400.

Moermond, T. C. and J. S. Denslow. 1985. Neotropical avian frugivores: Patterns of behavior, morphology, and nutrition with consequences for fruit selection. Pp 867–897 In: *Neotropical ornithology* (P. A. Buckley, M. S. Foster, E. S. Morton, R. S. Ridgely and F. G. Buckley, Eds). Ornith. Monog. 36, Allen Press, Lawrence.

Møller, A. P. 1988. Female choice selects for male sexual tail ornaments. *Nature* 332: 640–642.

Møller, A. P. 1991. Why mated songbirds sing so much: mate guarding and male announcement of mate fertility status. *Amer. Nat.* 138: 994–1014.

Møller, A. P. 1992. Sexual selection in the monogamous swallow *Hirundo rustica*. II. Mechanisms of intersexual selection. *J. Evol. Biol.* 5: 603–624.

Møller, A. P. 1994. *Sexual selection and the barn swallow.* Oxford Univ. Press, Oxford.

Møller, A. P. 1997. Immunce defence, extra-pair paternity, and sexual selection in birds. *Proc. R. Soc. Lond.* B 264: 561–566.

Møller, A. P. and T. R. Birkhead. 1994. The evolution of plumage brightness in birds is related to extra-pair paternity. *Evolution* 48: 1089–1100.

Møller, A. P. and J. V. Briskie. 1995. Extrapair paternity, sperm competition and the evolution of testis size in birds. *Behav. Ecol. Sociobiol.* 36: 357–365.

Møller, A. P., R. Dufva and J. Erritzøe. 1998a. Host immune function and sexual selection in birds. *J. Evol. Biol.* 11: 703–719.

Møller, A. P., P.-Y. Henry and J. Erritzøe. 2000. The evolution of song repertoires and immune defense in birds. *Proc. Biol. Soc. Lond.* 267: 1439.

Moore, O., B. J. M. Stutchbury and J. Quinn. 1999. Extra-pair mating system of an asynchronously breeding tropical songbird, the Mangrove Swallow. *Auk* 116: 1039–1046.

Moreau, R. E. 1937. Breeding seasons of birds in East African evergreen forest. *Proc. Zool. Soc. Lond.* **136**: 631–653.

Moreau, R. E. 1950. The breeding seasons of African birds. 1. Land birds. *Ibis* **92**: 223–267.

Morse, D. H. 1970. Ecological aspects of some mixed-species flocks of birds. *Ecol. Monogr.* **40**: 119–168.

Morse, D. H. 1974. Niche breadth as a function of social dominance. *Am. Nat.* **108**: 818–830.

Morse, D. H. 1977. Feeding behavior and predator avoidance in heterospecific groups. *BioScience* **27**: 332–339.

Morton, E. S. 1968. Robins feeding on hairy caterpillars. *Auk* **85**: 696.

Morton, E. S. 1971a. Food and migration habits of the eastern kingbird in Panama. *Auk* **88**: 925–926.

Morton, E. S. 1971b. Nest predation affecting the breeding season of the clay-colored robin, a tropical song bird. *Science* **181**: 920–921.

Morton, E. S. 1973. On the evolutionary advantages and disadvantages of fruit eating in tropical birds. *Amer. Nat.* **107**: 8–22.

Morton, E. S. 1977a. Intratropical migration in the yellow-green vireo and piratic flycatcher. *Auk* **94**: 97–106.

Morton, E. S. 1977b. On the occurrence and significance of motivation-structural rules in some bird and mammal sounds. *Amer. Nat.* **111**: 855–869.

Morton, E. S. 1978. Reintroducing recently extirpated birds into a tropical forest preserve. Pp 379–384 In: *Endangered birds: management techniques for preserving threatened species* (S. A. Temple, Ed.). Univ. of Wisconsin Press, Madison.

Morton, E. S. 1979a. A comparative survey of avian social systems in northern Venezuelan habitats. Pp 233–259 In: *Vertebrate ecology in the northern neotropics* (J. F. Eisenberg, Ed.). Smithsonian Institution Press, Washington.

Morton E. S. 1979b. Effective pollination of *Erythrina fusca* by the Orchard Oriole (*Icterus spurius*): coevolved behavioral manipulation? *Ann. Missouri Bot. Garden* **66**: 482–489.

Morton, E. S. 1980. Adaptations to seasonal changes by migrant land birds in the Panama Canal Zone. Pp 437–453 In: *Migrant birds in the neotropics: ecology, behavior, distribution, and conservation* (A. Keast and E. S. Morton, Eds). Smithsonian Institution Press, Washington.

Morton, E. S. 1982. Grading, discreteness, redundancy, and motivation-structural rules. Pp183–212 In: *Acoustic communication in birds*, vol. 1 (D. E. Kroodsma and T. E. Miller, Eds). Academic Press, New York.

Morton, E. S. 1983. Turdus grayi. Pp 610–611 In: *Costa Rican Natural History* (D. J. Janzen, Ed.). Univ. of Chicago Press, Chicago.

Morton, E. S. 1986. Predictions from the ranging hypothesis for the evolution of long distance signals in birds. *Behaviour* **99**: 65–86.

Morton, E. S. 1987. The effects of distance and isolation on song-type sharing in the Carolina Wren. *Wilson Bull.* **99**: 601–610.

Morton, E. S. 1992. What do we know about the future of migrant landbirds? Pp. 579–589 In: *Ecology and conservation of neotropical landbirds* (J. M. Hagen and D. W. Johnston, Eds). Smithsonian Institution Press, Washington.

Morton, E. S. 1996a. Why songbirds learn songs: an arms race over ranging? *Poultry and Avian Biology Reviews* **7**: 65–71.

Morton, E. S. 1996b. A comparison of vocal behavior among tropical and temperate passerine birds. Pp 258–268 In: *Ecology and evolution of acoustic communication in birds* (D. E. Kroodsma and E. H. Miller, Eds). Cornell Univ. Press, Ithaca.

Morton, E. S. and K. C. Derrickson. 1996. Song ranging by the dusky antbird, *Cercomacra tyrannina*: ranging without song learning. *Behav. Ecol. Sociobiol.* **39**: 195–201.

Morton, E. S. and J. Page. 1992. *Animal Talk: science and the voices of nature.* Random House, New York.

Morton, E. S. and M. D. Shalter. 1977. Vocal response to predators in pair-bonded Carolina Wrens. *Condor* **79**: 222–227.

Morton E. S. and B. J. M. Stutchbury. 2000. Demography and reproductive success in the dusky antbird, a sedentary tropical passerine. *J. Field Ornithol.* **71**: 493–500.

Morton, E. S., G. Geitgey, and S. McGrath. 1978. On bluebird 'responses to apparent female adultery.' *Amer. Nat.* **112**: 968–971.

Morton E. S., J. Lynch, K. Young and P. Mehlhop. 1986. Do male Hooded Warblers exclude females from nonbreeding territories in tropical forests? *Auk* **104**: 133–135.

Morton, E. S., L. Forman and M. Braun. 1990. Extra-pair fertilizations and the evolution of colonial breeding in purple martins. *Auk* **107**: 275–283.

Morton, E. S., B. J. M. Stutchbury, J. S. Howlett and W. H. Piper. 1998. Genetic monogamy in blue-headed vireos and a comparison with a sympatric vireo with extrapair paternity. *Behav. Ecol.* **9**: 515–524.

Morton, E. S., K. C. Derrickson and B. J. M. Stutchbury. 2000. Territory switching in a sedentary tropical passerine the dusky antbird *Cercomacra tyrannina*. *Behav. Ecol.*, in press.

Moynihan, M. 1962. The organization and probable evolution of some mixed species flocks of Neotropical birds. *Smithsonian Misc. Coll.* **143**: 1–140.

Moynihan, M. 1998. *The social regulation of competition and aggression: with a discussion of tactics and strategies.* Smithsonian Inst. Press, Washington.

Munn, C. A. 1985. Permanent canopy and understory flocks in Amazonia: species composition and population density. Pp 683–712. In: *Neotropical ornithology* (P. A. Buckley, M. S. Foster, E. S. Morton, R. S. Ridgely and R. G. Buckley, Eds). *Ornithological Monograph* 36. Allen Press, Kansas.

Munn, C. A. 1986. Birds that 'cry wolf'. *Nature* **319**: 143–145.

Munn, C. A. and J. W. Terborgh. 1979. Multi-species territoriality in neotropical foraging flocks. *Condor* **81**: 338–347.

Murray, B. G., Jr. 1985. Evolution of clutch size in tropical species of birds. *Ornithol. Monogr.* **36**: 505–519.

Murray, K. G., S. Russell, C. M. Picone, K. Winnett-Murray, W. Sherwood and M. L. Kuhlmann. 1994. Fruit laxatives and seed passage rates in frugivores: consequences for plant reproductive success. *Ecology* **75**: 989–994.

Naef-Daenzer, B. and L. F. Keller. 1999. The foraging performance of great and blue tits (*Parus major* and *P. caeruleus*) in relation to caterpillar development, and its consequences for nestling growth and fledgling weight. *J. Anim. Ecol.* **68**: 708–718.

Nager, R. G. and A. J. van Noordwijk. 1995. Proximate and ultimate aspects of phenotypic plasticity in timing of great tit breeding in a heterogenous environment. *Amer. Nat.* **146**: 454–474.

Naguib, M. 1995. Auditory distance assessment of singing conspecifics in Carolina wrens: the role of reverberation and frequency-dependent attenuation. *Anim. Behav.* **50**: 1297–1307.

Nealen, P. M. and D. J. Perkel. 2000. Sexual dimorphism in the song system of the Carolina wren *Thryothorus ludovicianus*. *J. Comp. Neurol.* **418**: 346–360.

Nee, S. and R. M. May. 1997. Extinction and the loss of evolutionary history. *Science* **278**: 692–694.

Negro, J. J., M. Villarroel, J. L. Tella, U. Kuhnleiu, F. Hilaldo, J. A. Donazar and D. M. Bird. 1996. DNA fingerprinting reveals low incidence of EPFs in the Lesser Kestrel. *Anim. Behav.* **51**: 935–943.

Neudorf, D. L., B. J. M. Stutchbury and W. H. Piper. 1997. Covert extraterritorial behavior of female hooded warblers. *Behav. Ecol.* **8**: 595–600.

Nice, M. M. 1937. Studies in the life history of the Song Sparrow-I. *Trans. Linn. Soc. N.Y.* **4**: 1–247.

Nilsson, J. A. and E. Svensson. 1993. Energy constraints and ultimate decisions during egg-laying in the blue tit. *Ecology* **74**: 244–251.

Norris, D. R. and B. J. M. Stutchbury. 2001. Extra-territorial movements of a forest songbird in a fragmented landscape. *Cons. Biol.* in press.

Nowicki, S., S. Peters and J. Podos. 1998. Song learning, early nutrition, and sexual selection in songbirds. *Amer. Zool.* **38**: 179–190.

Nur, N. 1988. The consequences of brood size for breeding blue tits. III. Measuring the cost of reproduction: survival, future fecundity, and differential dispersal. *Evolution* **42**: 351–362.

Nur, N. 1990. The cost of reproduction in birds: evaluating the evidence from manipulative and non-manipulative studes. Pp 281–296 In: *Population biology of passerine birds, an integrated approach* (J. Blondel, A. Gosler, J. D. Lebreton and R. McCleery, Eds). Springer-Verlag.

Ohala, J. J. 1984. An ethological perspective on common cross-language utilization of F$_0$ of voice. *Phonetica* **41**: 1–16.

Oniki, Y. 1979. Is nesting success of birds low in the Tropics? *Biotropica* **11**: 60–69.

Orians, G. H. 1969. On the evolution of mating systems in birds and mammals. *Amer. Nat.* **103**: 589–603.

Orians, G. H. 1985. *Blackbirds of the Americas*. Univ. of Washington Press,

Seattle.

Oring, L. W., A. J. Fivizzani and M. E. El Halawani. 1989. Testosterone-induced inhibition of incubation in the spotted sandpiper (*Actitis mecularia*). *Horm. Behav.* **23**: 412–423.

Otter, K., L. Ratcliffe and P. T. Boag. 1994. Extra-pair paternity in the Black-capped Chickadee. *Condor* **96**: 218–222.

Otter, K., B. Chruszcz and L. Ratcliffe. 1997. Honest advertisement and song output during the dawn chorus of black-capped chickadees. *Behav. Ecol.* **8**: 167–173.

Otter, K., L. Ratcliffe, D. Michaud and P. T. Boag. 1998. Do female black-capped chickadees prefer high-ranking males as extra-pair partners? *Behav. Ecol. Sociobiol.* **43**: 25–36.

Owings, D. H. and E. S. Morton. 1998. *Animal vocal communication: a new approach.* Cambridge Univ. Press, Cambridge.

Park, S. R. and D. Park. 2000. Song type for intrasexual interaction in the bush warbler. *Auk* **117**: 228–232.

Parrish, J. D. 2000. Behavioral, energetic, and conservation implications of foraging plasticity during migration. *Studies in Avian Biol.* **20**: 53–70.

Pärt, T. 1991. Is dawn singing related to paternity insurance? The case of the collared flycatcher. *Anim. Behav.* **41**: 451–456.

Payne, R. B. 1983. The social context of song mimicry: song matching dialects in Indigo Buntings (*Passerina cyanea*). *Anim. Behav.* **31**: 788–805.

Payne, R. B. 1996. Song traditions in indigo buntings: origin, improvisation, dispersal, and extinction in cultural evolution. Pp 198–221 In: *Ecology and evolution of acoustic communication in birds* (D. E. Kroodsma and E. H. Miller, Eds). Cornell University Press, Ithaca.

Peek, F. W. 1972. An experimental study of the territorial function of vocal and visual displays in the male red-winged blackbirds (*Agelaius phoeniceus*). *Anim. Behav.* **29**: 112–178.

Perrins, C. M. 1970. The timing of birds' breeding seasons. *Ibis* **112**: 242–255.

Perrins, C. M. 1991. Tits and their caterpillar food supply. *Ibis* **133**: 49–54.

Peters, W. D. and T. C. Grubb, Jr. 1983. An experimental analysis of sex-specific foraging in the Downy Woodpecker, *Picoides pubescens. Ecology* **64**: 1437–1443.

Petren, K., B. R. Grant and P. R. Grant. 1999. Low extrapair paternity in the Cactus Finch (*Geospiza scandens*). *Auk* **116**: 252–256.

Pitcher, T. E. and B. J. M. Stutchbury. 2000. Extraterritorial forays and male parental care in hooded warblers. *Anim. Behav.* **59**: 1261–1269.

Poulin, B., G. Lefebvre and R. McNeil. 1992. Tropical avian phenology in relation to abundance and expoitation of food resources. *Ecology* **73**: 2295–2309.

Poulin, R. 1996. Sexual inequalities in helminth infections: a cost of being male? *Amer. Nat.* **147**: 287–295.

Powell, G. V. N. 1979. Structure and dynamics of interspecific flocks in a Neotropical mid-elevation forest. *Auk* **96**: 375–390.

Powell, G. V. N. 1985. Sociobiology and adaptive significance of interspecific foraging flocks in the neotropics. Pp 713–732 In: *Neotropical ornithology* (P. A. Buckley, M. S. Foster, E. S. Morton, R. S. Ridgely and R. G. Buckley, Eds). *Ornithological Monograph* 36. Allen Press, Kansas.

Prum, R. O. and V. R. Razafindratsita. 1997. Lek behavior and natural history of the Velvet Asite (*Philepitta castanea*: Eurylaimidae). *Wilson Bull.* **109**: 371–392.

Purvis, A., P-M Agapow, J. L. Gittleman and G. M. Mace. 2000. Nonrandom extinction and the loss of evolutionary history. *Science* **288**: 328–330.

Quarström, A. 1997. Experimentally increased badge size increases male competition and reduces male parental care in the collared flycatcher. *Proc. Royal Soc. Lond.* B **264**: 1225–1231.

Quinn, J. S., G. E. Woolfenden, J. W. Fitzpatrick and B. N. White. 1999. Multi-locus fingerprinting supports monogamy in Florida scrub-jays. *Behav. Ecol. Sociobiol.* **45**: 1–10.

Rabenold, K. N. 1990. *Campylorhynchus* wrens: the ecology of delayed dispersal and cooperation in the Venezuelan savannah. Pp 157–196 In: *Cooperative breeding in birds: long term studies of ecology and behavior* (P. B. Stacey and W. D. Koenig, Eds). Cambridge Univ. Press, Cambridge.

Rabenold, P. P., K. N. Rabenold, W. H. Piper and S. W. Zack. 1990. Shared paternity revealed by genetic analysis in cooperatively breeding tropical wrens. *Nature* **348**: 538–540.

Rabøl, J. 1987. Coexistence and competition between overwintering Willow Warblers *Phylloscopus trochilus* and local warblers at Lake Naivasha, Kenya. *Ornis Scand.* **18**: 101–121.

Radesater, T., S. Jakobsson, N. Andbjer, A. Bylin and K. Nystrom. 1987. Song rate and pair formation in the willow warbler, *Phylloscopus trochilus*. *Anim. Behav.* **35**: 1645–1651.

Ralph, C. J. and S. G. Fancy. 1994. Timing of breeding and molting in six species of Hawaiian honeycreepers. *Condor* **96**: 151–161.

Ramsay, S. L. and D. C. Houston. 1997. Nutritional constraints on egg production in the blue tit: a supplementary feeding study. *J. Anim. Ecol.* **66**: 649–657.

Raouf, S. A., P. G. Parker, E. D. Ketterson, V. Nolan Jr., and C. Ziegenfus. 1997. Testosterone affects reproductive success by influencing extra-pair fertilizations in male dark-eyed juncos (Aves: *Junco hyemalis*). *Proc. R. Soc. Lond.* B **264**: 1599–1603.

Rappole, J. H. 1995. *The ecology of migrant birds, a Neotropical perspective*. Smithsonian Inst. Press, Washington, D. C.

Rappole, J. H. and E. S. Morton. 1985. Effects of habitat alteration on a tropical avian forest community. Pp 1013–1021 In: *Neotropical ornithology* (Buckley, P. A., M. S. Foster, E. S. Morton, R. S. Ridgely and R. G. Buckley, Eds). *Ornithological Monograph* 36. Allen Press, Kansas.

Rappole, J. H., E. S. Morton, T. E. Lovejoy III, and J. L. Ruos. 1983. *Nearctic avian migrants in the Neotropics*. USDI, Fish and Wildlife Service, Washington, D. C. 384 pp.

Ratcliffe, L. and K. Otter. 1996. Sex differences in song recognition. Pp 339–355 In: *Ecology and evolution of acoustic communication in birds* (D. E. Kroodsma and E. H. Miller, Eds). Cornell Univ. Press, Ithaca.

Regal, P. J. 1976. Ecology and evolution of flowering plant dominance. *Science* **196**: 622–629.

Reid, M. L. 1987. Costliness and reliability in the singing vigour of Ipswich sparrow. *Anim. Behav.* **35**: 1735–1743.

Remsen, J. V., Jr. 1984. High incidence of 'leapfrog' pattern of geographic variation in Andean birds: implications for the speciation process. *Science* **224**: 171–172.

Restrepo, C. and M. L. Mondragón. 1998. Cooperative breeding in the frugivorous Toucan Barbet (*Semnornis ramphastinus*). *Auk* **115**: 4–15.

Reyer, H. U. 1990. Pied Kingfishers: ecological causes and reproductive consequences of cooperative breeding. Pp 527–558 In: *Cooperative breeding in birds: long term studies of ecology and behavior* (P. B. Stacey and W. D. Koenig, Eds). Cambridge Univ. Press, Cambridge.

Richards, D. G. 1981. Estimation of distance of singing conspecifics by the Carolina wren. *Auk* **98**: 127–133.

Richardson, K. C. and R. D. Wooller. 1988. The alimentary tract of a specialist frugivore, the mistletoebird, *Dicaeum hirundinaceum*, in relation to its diet. *Australian J. Zool.* **36**: 373–382.

Ricklefs, R. 1966. The temporal component of diversity among species of birds. *Evolution* **20**: 235–242.

Ricklefs, R. E. 1969a. An analysis of nesting mortality in birds. *Smithsonian Contr. Zool.* **9**: 1–48.

Ricklefs, R. 1969b. The nesting cycle of songbirds in tropical and temperate regions. *Living Bird* **8**: 1–48.

Ricklefs, R. E. 1977. On the evolution of reproductive strategies in birds: reproductive effort. *Amer. Nat.* **111**: 453–478.

Ricklefs, R. E. 1980. Geographical variation in clutch size among passerine birds: Ashmole's hypothesis. *Auk* **97**: 38–49.

Ricklefs, R. 1997. Comparative demography of New World populations of thrushes (*Turdus* spp.). *Ecol. Monogr.* **67**: 23–43.

Ridgely, R. S. and J. A. Gwynne. 1989. *A guide to the birds of Panama.* 2nd ed. Princeton Univ. Press, Princeton.

Ritchison, G. 1983. The function of singing in female black-headed grosbeaks: family-group maintenance. *Auk* **100**: 105–116.

Robbins, M. B., G. H. Rosenberg and F. S. Molina. 1994. A new species of cotinga (Cotingidae: *Doliornis*) from the Ecuadorian Andes, with comments on plumage sequences in *Doliornis* and *Ampelion. Auk* **111**: 1–7.

Robertson, B. C. 1996. The mating system of the Capricorn silvereye. Ph.D. Dissertation. University of Queensland, St. Lucia, Brisbane.

Robinson, D. 1990. The nesting ecology of sympatric Scarlet Robin *Petroica multicolor* and Flame Robin *P. phoenicea* populations in open Eucalypt forest. *Emu* **90**: 40–52.

Robinson, S. K. 1985. Coloniality in the yellow-rumped cacique as a defense against nest predators. *Auk* **102**: 506–519.

Robinson, S. K. 1986. Competitive and mutualistic interactions among females in a neotropical oriole. *Anim. Behav.* **34**: 113–122.

Robinson, S. K. and J. Terborgh. 1995. Interspecific aggression and habitat selection by Amazonian birds. *J. Anim. Ecol.* **64**: 1–11.

Robinson, S. K., F. R. Thompson III, T. M. Donovan, D. R. Whitehead and J. Faaborg. 1995. Regional forest fragmentation and the nesting success of migratory birds. *Science* **267**: 1987–1990.

Robinson, T. R., W. D. Robinson and E. C. Edwards. 2000. Breeding ecology and nest-site selection of Song Wrens in central Panama. *Auk* **117**: 345–354.

Robinson, W. D., T. R. Robinson, S. K. Robinson and J. D. Brawn. 2000. Nesting success of understory forest birds in central Panama. *J. Avian Biol.* **31**: 151–164.

Rohwer, S. 1982. The evolution of reliable and unreliable badges of fighting ability. *Amer. Zool.* **22**: 531–546.

Rohwer, S. and G. S. Butcher. 1988. Winter versus summer explanations of delayed plumage maturation in passerine birds. *Amer. Nat.* **131**: 556–572.

Rohwer, S. and E. Roskaft. 1989. Results of dyeing male yellow-headed blackbirds solid black: implications for the arbitrary identity badge hypothesis. *Behav. Ecol. Sociobiol.* **25**: 39–48.

Rohwer, S., S. D. Fretwell and D. M. Niles. 1980. Delayed maturation in passerine plumages and the deceptive acquistion of resources. *Amer. Nat.* **115**: 400–437.

Roper, J. J. and R. R. Goldstein. 1997. A test of the Skutch hypothesis: does activity at nests increase nest predation risk? *J. Avian Biol.* **28**: 111–117.

Rowan, M. K. 1966. Territory as a density-regulating mechanism in some South African birds. *Ostrich* (suppl.) **6**: 397–408.

Rowley, I. and E. Russell. 1991. Demography of passerines in the temperate southern hemisphere. Pp 22–44 In: *Bird population studies: relevance to conservation, demography and management* (C. M. Perrins, J. D. Lebreton and G. J. M. Hirons, Eds). Oxford Univ. Press, Oxford.

Saino, N., A. M. Bolzern and A. P. Møller. 1997a. Immunocompetence, ornamentation, and viability of male barn swallows (*Hirundo rustica*). *Proc. Natl. Acad. Sci. USA* **94**: 549–552.

Saino, N., C. R. Primmer, H. Ellegren and A. P. Møller. 1997b. An experimental study of paternity and tail ornamentation in the barn swallow (*Hirundo rustica*). *Evolution* **51**: 562–570.

Sandercock, B. K., S. R. Beissinger, S. H. Stoleson, R. R. Melland and C. R. Hughes. 2000. Survival rates of a neotropical parrot: implications for latitudinal comparisons of avian demography. *Ecology* **81**: 1351–1370.

Sargent, S. 1993. Nesting biology of the Yellow-throated Euphonia: large clutch size in a neotropical frugivore. *Wilson Bull.* **105**: 285–300.

Schneider, K. J. 1984. Dominance, predation, and optimal foraging in White-throated Sparrow flocks. *Ecology* **65**: 1820–1827.

Schoech, S. J. 1996. The effect of supplemental food on body condition and the timing of reproduction in a cooperative breeder, the Florida scrub jay. *Condor* **98**: 234–244.

Schwagmeyer, P. L. and E. D. Ketterson. 1999. Breeding synchrony and EPF rates: the key to a can of worms? *Trends Ecol. Evol.* **14**: 47–48.

Schwagmeyer, P. L., R. C. St. Clair, J. D. Moodie, T. C. Lamey, G. D. Schnell and M. N. Moodie. 1999. Species differences in male parental care in birds: a reexamination of correlates with paternity. *Auk* **116**: 487–503.

Sealy, S. G. 1989. Defence of nectar sources by migrating Cape May Warblers. *J. Field Ornithol.* **60**: 89–93.

Searcy, W. A. and K. Yasukawa. 1996. Song and female choice. Pp 454–473 In: *Ecology and evolution of acoustic communication in birds* (D. E. Kroodsma, and E. H. Miller, Eds). Cornell Univ. Press, Ithaca.

Selander, R. K. 1965. On mating systems and sexual selection. *Amer. Nat.* **99**: 129–141.

Sheldon, B. C. 1994. Sperm competition in the chaffinch: the role of the female. *Anim. Behav.* **47**: 163–173.

Sick, H. 1993. *Birds in Brazil*. Princeton U. Press.

Sieving, K. E. 1992. Nest predation and differential insular extinction among selected forest birds of central Panama. *Ecology* **73**: 2310–2328.

Simmons, R. E. 1993. Effects of supplementary food on density-reduced breeding in an African eagle: adaptive restraint or ecological constraint? *Ibis* **135**: 394–402.

Sinclair, A. R. E. 1978. Factors affecting the food supply and breeding season of resident birds and movements of Palearctic migrants in a tropical African savannah. *Ibis* **120**: 480–497.

Skutch, A. F. 1935. Helpers at the nest. *Auk* **52**: 257–273.

Skutch, A. F. 1949. Do tropical birds rear as many young as they can nourish? *Ibis* **91**: 430–455.

Skutch, A. F. 1950. The nesting seasons of Central American birds in relation to climate and food supply. *Ibis* **92**: 185–222.

Skutch, A. F. 1954. *Life Histories of Central American Birds*. Pacific Coast Avifauna No. 31, Cooper Ornithological Society, Berkeley California.

Skutch, A. F. 1960. Life histories of Central American birds. *Pacific Coast Avifauna* **34**: 1–593.

Skutch, A. F. 1969. Life histories of Central American birds. *Pacific Coast Avifauna* **35**: 1–580.

Skutch, A. F. 1981. *New studies of tropical American birds*. Publ. Nuttall Ornithol. Club No. 19.

Skutch, A. F. 1985. Clutch size, nesting success, and predation on nests of neotropical birds, reviewed. *Ornithol. Monogr.* **36**: 575–594.

Skutch, A. F. 1996. *Antbirds and ovenbirds*. Univ. of Texas Press, Austin.

Skutch, A. F. 1997. *Life of the Flycatcher*. Univ. of Oklahoma Press, Oklahoma.

Slagsvold, T. and J. Lifjeld. 1997. Incomplete female knowledge may explain variation in extra-pair paternity in birds. *Behavior* **134**: 1–19.

Smith, D. G. 1972. The role of the epaulets in the Red-winged Blackbird social system (*Agelaius phoeniceus*). *Behavior* **41**: 251–268.

Smith, S. M. 1978. The 'underworld' in a territorial sparrow: adaptive strategy for floaters. *Amer. Nat.* **112**: 571–582.

Smith, S. M. 1988. Extra-pair copulations in Black-capped Chickadees: the role of the female. *Behaviour* **107**: 15–23.

Smith, S. M. 1991. *The Black-capped Chickadee*. Cornell University Press, Ithaca.

Snow, B. K. 1970. A field study of the bearded bellbird in Trinidad. *Ibis* **112**: 299–329.

Snow, B. K. 1974. Lek behaviour and breeding of Guy's Hermit Humming-bird *Phaethornis guy. Ibis* **116**: 278–297.

Snow, B. K. and D. W. Snow. 1979. The Ochre-bellied Flycatcher and the evolution of lek behavior. *Condor* **81**: 286–292.

Snow, B. K. and D. W. Snow. 1988. *Birds and Berries.* T & A D Poyser, Calton.

Snow, D. W. 1962. A field study of the Black and White Manakin, *Manacus manacus*, in Trinidad. *Zoologica* **47**: 65–104.

Snow, D. W. 1965. A possible selective factor in the evolution of fruiting seasons in tropical forest. *Oikos* **15**: 274–281.

Snow, D. W. 1971. Evolutionary aspects of fruit-eating by birds. *Ibis* **113**: 194–202.

Snow, D. W. 1976a. The relationship between climate and annual cycles in the Cotingidae. *Ibis* **118**: 366–401.

Snow, D. W. 1981. Tropical frugivorous birds and their food plants: a world survey. *Biotropica* **13**: 1–14.

Snow, D. W. and A. Lill. 1974. Longevity records for some Neotropical land-birds. *Condor* **76**: 262–267.

Snow, D. W. and B. K. Snow. 1963. Breeding and the annual cycle in three Trinidad thrushes. *Wilson Bull.* **75**: 27–41.

Snow, D. W. and B. K. Snow. 1973. The breeding of the Hairy Hermit *Glaucis hirsuta* in Trinidad. *Ardea* **61**: 106–122.

Soler, M., M. Martin-Vivaldi, J. M. Marin and A. P. Møller. 1999. Weight lifting and health status in the black wheatear. *Behav. Ecol.* **10**: 281–286.

Sorenson, L. G. 1992. Variable mating system of a sedentary tropical duck: the White-cheeked Pintail (*Anas bahamensis bahamensis*). *Auk* **109**: 277–292.

Stacey, P. B. and J. D. Ligon. 1991. The benefits-of-philopatry hypothesis for the evolution of cooperative breeding: variation in territory quality and group size effects. *Amer. Nat.* **137**: 831–846.

Staicer, C. A. 1992. Social behavior of the Northern Parula, Cape May Warbler, and Prairie Warbler wintering in second-growth forest in south-western Puerto Rico. Pp 309–320 In: *Ecology and conservation of neotropical migrant landbirds* (J. M. Hagan III and D. W. Johnston, Eds). Smithsonian Inst. Press, Washington.

Staicer, C. A., D. A. Spector and A. G. Horn. 1996. The dawn chorus and other diel patterns in acoustic signaling. Pp 426–453 In: *Ecology and evolution of acoustic communication in birds* (D. E. Kroodsma and E. H. Miller, Eds). Cornell Univ. Press, Ithaca.

Stiles, E. W. 1980. Patterns of fruit presentation and seed dispersal in bird-disseminated woody plants in the eastern deciduous forest. *Amer. Nat.* **116**: 670–688.

Stiles, F. G. 1980. The annual cycle in a tropical wet forest hummingbird community. *Ibis* **122**: 322–343.

Stiles, F. G. 1992. Effects of a severe drought on the population biology of a tropical hummingbird. *Ecology* **73**: 1375–1390.

Stiles, F. G. and L. L. Wolf. 1974. A possible circannual molt rhythm in a tropical hummingbird. *Amer. Nat.* **108**: 341–354.

Stiles, F. G. and L. L. Wolf. 1979. Ecology and evolution of lek mating behavior in the long-tailed hermit hummingbird. *Ornithol. Monogr.* 27. American Ornithol. Union.

Stokes, A. W. 1974. *Territory. Benchmark papers in animal behavior.* Dowden, Hutchinson & Ross, Inc., Stroudsburg, PA.

Stoleson, S. H. and S. R. Beissinger. 1997. Hatching asynchrony, brood reduction, and food limitation in a neotropical parrot. *Ecol. Monogr.* **67**: 131–154.

Stouffer, P. C. 1997. Interspecific aggression in *Formicarius* antthrushes? The view from central Amazonian Brazil. *Auk* **114**: 780–785.

Strahl, S. D. and A. Schmitz. 1990. Hoatzins: cooperative breeding in a folivorous neotropical bird. Pp 131–156 In: *Cooperative breeding in birds: long term studies of ecology and behavior* (P. B. Stacey and W. D. Koenig, Eds). Cambridge Univ. Press, Cambridge.

Strain, J. G. and R. L. Mumme. 1988. Effects of food supplementation, song playback, and temperature on vocal territorial behavior of Carolina Wrens. *Auk* **105**: 11–16.

Stratford, J. A. and P. C. Stouffer. 1999. Local extinctions of terrestrial insectivorous birds in a fragmented landscape near Manaus, Brazil. *Cons. Biol.* **13**: 1416–1423.

Studd, M. V. and R. J. Robertson. 1985. Life span, competition, and delayed plumage maturation in male passerines: the breeding threshold hypothesis. *Amer. Nat.* **126**: 101–115.

Stutchbury, B. J. 1991. The adaptive significance of male subadult plumage in purple martins: plumage dyeing experiments. *Behav. Ecol. Sociobiol.* **29**: 297–306.

Stutchbury, B. J. 1992. Experimental evidence that bright coloration is not important for territory defense in purple martins. *Behav. Ecol. Sociobiol.* **31**: 27–33.

Stutchbury, B.J. 1994. Competition for winter territories in a Neotropical migrant songbird: the role of age, sex, and color. *Auk* **111**: 63–69.

Stutchbury, B. J. M. 1998a. Female mate choice of extra-pair males: breeding synchrony is important. *Behav. Ecol. Sociobiol.* **43**: 213–215.

Stutchbury, B. J. M. 1998b. Extra-pair mating effort of male hooded warblers, *Wilsonia citrina. Anim. Behav.* **55**: 553–561.

Stutchbury, B. J. M. and J. S. Howlett. 1995. Does male-like coloration in female Hooded Warblers increase nest predation? *Condor* **97**: 559–564.

Stutchbury, B. J. and E. S. Morton. 1995. The effect of breeding synchrony on extra-pair mating systems in songbirds. *Behaviour* **132**: 675–690.

Stutchbury, B. J. M. and D. Neudorf. 1998. Female control, breeding synchrony and the evolution of extra-pair mating strategies. Pp 103–122 In: Avian reproductive tactics: female and male perspectives, (P. Parker and N. Burley, Eds.). *Ornithol. Monogr.* **49**, Amer. Ornithol. Union.

Stutchbury, B. J. and R. J. Robertson. 1987a. Behavioral tactics of subadult female floaters in the tree swallow. *Behav. Ecol. Sociobiol.* **20**: 413–419.

Stutchbury, B. J. and R. J. Robertson. 1987b. Signaling subordinate and female status: two hypotheses for the adaptive significance of subadult plumage in female tree swallows. *Auk* **104**: 717–723.

Stutchbury, B. J., J. M. Rhymer, and E. S. Morton. 1994. Extra-pair paternity in the hooded warbler. *Behav. Ecol.* **5**: 384–392.

Stutchbury, B. J. M., W. H. Piper, D. L. Neudorf, S. A. Tarof, J. M. Rhymer, G. Fuller and R. C. Fleischer. 1997. Correlates of extra-pair fertilization success in hooded warblers. *Behav. Ecol. Sociobiol.* **40**: 119–126.

Stutchbury, B. J. M., E. S. Morton and W. H. Piper. 1998. Extra-pair mating system of a synchronously breeding tropical songbird. *J. Avian Biol.* **29**: 72–78.

Sullivan, K. A. 1984. The advantages of social foraging in Downy Woodpeckers. *Anim. Behav.* **32**: 16–22.

Svensson, E. 1997. Natural selection on avian breeding time: causality, fecundity-dependent, and fecundity-independent selection. *Evolution* **51**: 1276–1283.

Svensson, E. and J. A. Nilsson. 1995. Food supply, territory quality, and reproductive timing in the blue tit (*Parus caeruleus*). *Ecology* **76**: 1804–1812.

Tallman, D. A. and E. J. Tallman. 1997. Timing of breeding by antbirds (Formicariidae) in an aseasonal environment in Amazonian Ecuador. *Ornithol. Monogr.* **48**: 783–789.

Tarburton, M. K. 1987. An experimental manipulation of clutch and brood size of white-rumped swiftlets *Aerodramus spodiopygius* of Fiji. *Ibis* **129**: 107–114.

Tarof, S. A., B. J. M. Stutchbury, W. H. Piper and R. C. Fleischer. 1998. Does high breeding density increase the frequency of extra-pair fertilizations in Hooded Warblers? *J. Avian Biol.* **29**: 145–154.

Temeles, E. J., I. L. Pan, J. L. Brennan and J. N. Horwitt. 2000. Evidence for ecological causation of sexual dimorphism in a hummingbird. *Science* **289**: 441–443.

Terborgh J. 1990. Mixed flocks and polyspecific associations: costs and benefits of mixed groups to birds and monkeys. *Amer. J. Primatol.* **21**: 87–100.

Terborgh, J. S., S. K. Robinson, T. A. Parker III, C. A. Munn and N. Pierpont. 1990. Structure and organization of an Amazonian forest bird community. *Ecol. Monogr.* **60**: 213–238.

Thiollay, J. 1988. Comparative foraging success of insectivorous birds in tropical and temperate forests: ecological implications. *Oikos* **53**: 17–30.

Thiollay, J. 1994. Structure, density and rarity in an Amazonian rainforest bird community. *J. Trop. Ecol.* **10**: 449–481.

Thiollay, J. 1999. Frequency of mixed species flocking in tropical forest birds and correlates of predation risk: an intertropical comparison. *J. Avian Biol.* **30**: 282–294.

Thompson, J. N. and M. F. Willson. 1979. Evolution of temperate bird/fruit interactions: phenological strategies. *Evolution* **33**: 973–982.

Titus, R. C., C. R. Chandler, E. D. Ketterson and V. Nolan Jr. 1997. Song rates of dark-eyed juncos do not increase when females are fertile. *Behav. Ecol. Sociobiol.* **41**: 165–169.

Trail, P. W. 1985. Courtship disruption modifies mate choice in a lek-breeding bird. *Science* **227**: 778–779.

Trail, P. W. 1990. Why should lek-breeders be monomorphic? *Evolution* **44**: 1837–1852.

Trail, P. W. and E. S. Adams. 1989. Active mate choice at cock-of-the-rock leks: tactics of sampling and comparison. *Behav. Ecol. Sociobiol.* **25**: 283–292.

Trivers, R. 1972. Parental investment and sexual selection. Pp. 136–179 In: *Sexual selection and the descent of man, 1871–1971* (B. Campbell, Ed). Aldine, Chicago.

Tye, H. 1991. Reversal of breeding season by lowland birds at higher altitudes in western Cameroon. *Ibis* **134**: 154–163.

Vanderwerf, E. 1992. Lack's clutch size hypothesis: an examination of the evidence using meta-analysis. *Ecology* **73**: 1363–1374.

van Noordwijk, A. J., R. H. McCleery and C. M. Perrins. 1995. Selection for the timing of great tit breeding in relation to caterpillar growth and temperature. *J. Anim. Ecol.* **64**: 451–458.

Verhencamp, S. L. 1978. The adaptive significance of communal nesting in Groove-billed Anis (*Crotophaga sulcirostris*). *Behav. Ecol. Sociobiol.* **4**: 1–33.

Verhulst, S., S. J. Dieleman and H. K. Parmentier. 1999. A tradeoff between immunocompetence and sexual ornamentation in domestic fowl. *Proc. Natl. Acad. Sci. USA* **96**: 4478–4481.

Verner, J. and M. F. Willson. 1966. The influence of habitats on mating systems of North American passerine birds. *Ecology* **47**: 143–147.

Vleck, C. M. and J. L. Brown. 1999. Testosterone and social and reproductive behaviour in *Aphelocoma* jays. *Anim. Behav.* **58**: 943–951.

Wagner, R. H. 1993. The pursuit of extra-pair copulation by female birds: a new hypothesis of colony formation. *J. Theor. Biol.* **163**: 333–346.

Wagner, R. H. 1998. Hidden leks: sexual selection and the clustering of avian territories. *Ornithol. Monogr.* **49**: 123–145.

Wagner, R. H., M. D. Schug and E. S. Morton. 1996. Condition-dependent control of paternity by female purple martins: implications for coloniality. *Behav. Ecol. Sociobiol.* **38**: 379–389.

Weatherhead, P. J. and S. M. Yezerinac. 1998. Breeding synchrony and extra-pair matings in birds. *Behav. Ecol. Sociobiol.* **40**: 151–158.

Webster, M. S. 1995. Effects of female choice and copulations away from the colony on fertilization success of male Montezuma Oropendolas (*Psarocolius montezuma*). *Auk* 112: 659–671.

Webster, M. S. and S. K. Robinson. 1999. Courtship disruptions and male mating strategies: examples from female-defense mating systems. *Amer. Nat.* 154: 717–729.

Werner, T. K. and T. W. Sherry. 1987. Behavioral feeding specialization in *Pinaroloxias inornata*, the 'Darwin's Finch' of Cocos Island. *Proc. Natl. Acad. Sci. USA* 84: 5506–5510.

West-Eberhardt, M. J. 1983. Sexual selection, social competition and evolution. *Q. Rev. Biol.* 58: 155–183.

Westcott, D. 1997. Lek locations and patterns of female movement and distribution in a Neotropical frugivorous bird. *Anim. Behav.* 53: 235–247.

Westneat, D. F. and P. W. Sherman. 1993. Parentage and the evolution of parental behavior. *Behav. Ecol.* 4: 66–77.

Westneat, D. F. and P. W. Sherman. 1997. Density and extra-pair fertilizations in birds: a comparative analysis. *Behav. Ecol. Sociobiol.* 41: 205–215.

Westneat, D. F, P. W. Sherman and M. L. Morton. 1990. The ecology and evolution of extra-pair copulations in birds. *Curr. Ornithol.* 7: 331–369.

Wetmore, A. 1972. *Birds of the Republic of Panama. Part 3.* Smithsonian Inst. Press, Washington.

Wetmore, A. 1984. *Birds of the Republic of Panama. Part 4.* Smithsonian Inst. Press, Washington.

Wheelwright, N. T. 1986. The diet of American Robins: an analysis of the U.S. Biological Survey records. *Auk* 103: 710–725.

Wheelwright, N. T. 1988. Fruit-eating birds and bird-dispersed plants in the tropics and temperate zone. *Trends Ecol. Evol.* 3: 270–274.

Whitney, B. M. and J. Alvarez Alonso. 1998. A new *Herpsilochmus* Antwren (Aves: Thamnophilidae) from northern Amazonian Peru and adjacent Ecuador: the role of edaphic heterogeneity of *terra firme* forest. *Auk* 115: 559–576.

Whittingham, L. A., P. D. Taylor and R. J. Robertson. 1992a. Confidence of paternity and male parental care. *Amer. Nat.* 139: 1115–1125.

Whittingham, L. A., A. Kirkconnell and L. M. Ratcliffe. 1992b. Differences in song and sexual dimorphism between Cuban and North American Red-winged Blackbirds (*Agelaius phoeniceus*). *Auk* 109: 928–933.

Wikelski, M., M. Hau and J. C. Wingfield. 1999a. Social instability increases plasma testosterone in a year-round territorial neotropical bird. *Proc. Royal. Soc. Lond.* B 266: 551–556.

Wikelski, M., M. Hau, W. D. Robinson and J. C. Wingfield. 1999b. Seasonal endocrinology of tropical passerines: A comparative approach. Pp 1224–1241 In: *Proceedings of the 22nd Int. Ornithol. Congress* (N. J. Adams and R. H. Soltow, Eds) Birdlife South Africa, Johannesburg.

Wikelski, M., M. Hau, and J. C. Wingfield. 2000. Seasonality of reproduction in a neotropical rainforest bird. Ecology, 81: 2458–2472.

Wiley, R. H. 1991. Associations of song properties with habitats for territorial oscine birds of eastern North America. *Amer. Nat.* **138**: 973–993.

Wiley, R. H. and M. S. Wiley. 1977. Recognition of neighbors' duets by Stripe-backed Wrens *Campylorhynchus nuchalis. Behavior* **62**: 10–34.

Wiley, R. H., R. Godard and A. D. Thompson. 1994. Use of two singing modes by Hooded Warblers as adaptations for signalling. *Behavior* **129**: 243–278.

Wilkinson, R. 1983. Biannual breeding and moult-breeding overlap of the Chestnut-bellied Starling *Spreo pulcher. Ibis* **125**: 353–361.

Williams, M. and F. McKinney. 1996. Long term monogamy in a river specialist – the Blue Duck. Pp 73–90 In: *Partnerships in birds: the ecology of monogamy* (J.M. Black, Ed.). Oxford Univ. Press, Oxford.

Willis, E. O. 1966. Competitive exclusion and birds at fruiting trees in western Colombia. *Auk* **83**: 479–480.

Willis, E. O. 1967. The behavior of Bicolored Antbirds. *Univ. Calif. Publ. Zool.* **79**: 1–127.

Willis, E. O. 1972. The behavior of Spotted Antbirds. *Ornithol. Monogr.* **10**: 1–162.

Willis, E. O. 1973. The behavior of Oscellated Antbirds. *Smithsonian Contr. Zool.* **144**: 1–57.

Willis, E. O. 1974. Populations and local extinctions of birds on Barro Colorado Island, Panama. *Ecol. Monogr.* **44**: 153–169.

Willis, E. O. 1983. Longevities of some Panamanian forest birds, with note of low survivorship in old Spotted Antbirds (*Hylophylax naevioides*). *J. Field Ornithol.* **54**: 413–414.

Willis, E. O. 1985. *Cercomacra* and related antbirds (Aves, Formicariidae) as army ant followers. *Revta. Bras. Zool.* **2**: 427–432.

Willis, E. O. and Y. Oniki. 1978. Birds and army ants. *Ann. Rev. Ecol. Syst.* **9**: 243–263.

Willson, M. F. 1983. Avian frugivory and seed dispersal in eastern North America. *Curr. Ornithol.* **3**: 223–279.

Wilson, G. R. and M. C. Brittingham. 1998. How well do artificial nests estimate success of real nests? *Condor* **100**: 357–364.

Wingfield, J. C. 1984. Androgens and mating systems: testosterone-induced polygyny in normally monogamous birds. *Auk* **101**: 665–671.

Wingfield, J. C. 1994. Regulation of territorial behavior in the sedentary song sparrow, *Melospiza melodia morphna. Horm. Behav.* **28**: 1–15.

Wingfield, J. C. and T. P. Hahn. 1994. Testosterone and territorial behaviour in sedentary and migratory sparrows. *Anim. Behav.* **47**: 77–89.

Wingfield, J. C. and D. M. Lewis. 1993. Hormonal and behavioral responses to simulated territorial intrusion in the cooperatively breeding white-browed sparrow weaver, *Plocepasser mahali. Anim. Behav.* **45**: 1–11.

Wingfield, J. C. and M. C. Moore. 1987. Hormonal, social, and environmental factors in the reproductive biology of free-living male birds. Pp 148–175 In: *Psychobiology of reproductive behavior: an evolutionary perspective* (D. Crews, Ed.). Prentice-Hall, Inc., New Jersey.

Wingfield, J. C., R. E. Hegner Jr, A. M. Dufty Jr. and G. F. Ball. 1990. The 'challenge hypothesis': theoretical implications for patterns of testosterone secretion, mating systems, and breeding strategies. *Amer. Nat.* **136**: 829–846.

Wingfield, J. C., R. E. Hegner and D. M. Lewis. 1992. Hormonal responses to removal of a breeding male in the cooperatively breeding white-browed sparrow weaver, *Plocepasser mahali. Horm. Behav.* **26**: 145–155.

Wingfield, J. C., J. D. Jacobs, K. Soma, D. L. Maney, K. Hunt, D. Wisti-Peterson, S. Meddle, M. Ramenofsky and K. Sullivan. 1999. Testosterone, aggression, and communication: ecological bases of endocrine phenomena. Pp 255–283 In: *The Design of Animal Communication* (M. D. Hauser and M. Konishi, Eds). MIT Press, Cambridge, Massachusetts.

Witmer, M. C. 1993. Cooperative breeding by Rufous Hornbills on Mindanao Island, Philippines. *Auk* **110**: 933–936.

Wittenberger, J. F. 1979. A model for delayed reproduction in iteroparous animals. *Amer. Nat.* **114**: 439–446.

Wittenberger, J. L. and R. L. Tilson. 1980. The evolution of monogamy: hypotheses and evidence. *Ann. Rev. Ecol. and Syst.* **11**: 197–232.

Wolf, L., E. D. Ketterson and V. Nolan Jr. 1988. Paternal influence on growth and survival of dark-eyed junco young: Do parental males benefit? *Anim. Behav.* **36**: 1601–1618.

Wolf, L. L. 1969. Female territoriality in a tropical hummingbird. *Auk* **86**: 490–504.

Wolf, L. L. 1975. Female territoriality in the purple-throated carib. *Auk* **92**: 511–522.

Woodall, P. F. 1994. Breeding season and clutch size of the noisy pitta *Pitta versicolor* in tropical and subtropical Australia. *Emu* **94**: 273–277.

Woodworth, B. L. 1997. Brood parasitism, nest predation, and season-long reproductive success of a tropical island endemic. *Condor* **99**: 605–621.

Woodworth, B. L., J. Faaborg and W. J. Arendt. 1999. Survival and longevity of the Puerto Rican Vireo. *Wilson Bull.* **111**: 376–380.

Wrege, P. H. and S. T. Emlen. 1991. Breeding seasonality and reproductive success of White-fronted Bee-eaters in Kenya. *Auk* **108**: 673–687.

Wright, J. 1998. Paternity and paternal care. Pp 117–145 In: *Sperm competition and sexual selection* (T. R. Birkhead and A. P. Møller, Eds). Academic Press, London.

Wunderle, J. M. Jr. 1982. The timing of the breeding season in the bananaquit (*Coereba flaveola*) on the island of Grenada, W. I. *Biotropica* **14**: 124–131.

Wunderle, J. M. Jr. and K. H. Pollock. 1985. The banaquit-wasp nesting association and a random choice model. *Ornithol. Monogr.* **36**: 595–603.

Wyndham, E. 1986. Length of birds' breeding season. *Amer. Nat.* **128**: 155–164.

Yom-Tov, Y., M. I. Christie and G. J. Iglesia. 1994. Clutch size in passerines of southern South America. *Condor* **96**: 170–177.

Young, B. E. 1994. The effects of food, nest predation and weather on the timing of breeding in tropical house wrens. *Condor* **96**: 341–353.

Young, B.E. 1996. An experimental analysis of small clutch size in tropical house wrens. *Ecology* 77: 472–488.

Zack, S. 1990. Coupling delayed breeding with short-distance dispersal in cooperatively breeding birds. *Ethology* **86**: 265–286.

Zack, S. and J. D. Ligon. 1985a. Cooperative breeding in Lanius shrikes. I. Habitat and demography of two sympatic species. *Auk* 102: 754–765.

Zack, S. and J. D. Ligon. 1985b. Cooperative breeding in Lanius shrikes. II. Maintenance of group-living in a nonsaturated habitat. *Auk* **102**: 766–773.

Zack, S. and K. N. Rabenold. 1989. Assessment, age and proximity in dispersal contests among cooperatively breeding wrens: field experiments. *Anim. Behav.* **38**: 235–247.

Zack, S, and B. J. Stutchbury. 1992. Delayed breeding in avian social systems: the role of territory quality and 'floater' tactics. *Behavior* **123**: 194–219.

Zahavi, A. 1975. Mate selection – a selection for handicap. *J. Theor. Biol.* **53**: 205–214.

Zahavi, A. 1977. The cost of honesty (further remarks on the handicap principle). *J. Theor. Biol.* **67**: 603–605.

Zahavi, A. and A. Zahavi. 1997. *The handicap principle.* Oxford Univ. Press, Oxford.

Zandt, H., A. Strijkstra, J. Blondel and J. H. Van Balen. 1990. Food in two Mediterranean blue tit populations: do differences in caterpillar availability explain differences in timing of the breeding season. Pp 145–155 In: *Population biology of passerine birds, an integrated approach* (J. Blondel, A. Gosler, J. D. Lebreton and R. McCleery, Ed.). Berlin: Springer Verlag.

Index